CNC Tips and Techniques

A Reader for Programmers

Peter Smid

Industrial Press Inc.
New York

A full catalog record for this book will be available
from the Library of Congress.
ISBN: 978-0-8311-3472-3

Industrial Press, Inc.
989 Avenue of the Americas
New York, NY 10018

Sponsoring Editor: John Carleo
Developmental Editor: Robert Weinstein
Interior Text and Cover Design: Janet Romano
Cover Photo courtesy of: James A. Harvey

Copyright © 2013 by Industrial Press
Printed in the United States of America.
All rights reserved.
This book, or any parts thereof, may not be reproduced, stored in a retrieval system, or transmitted in any form without the permission of the publisher.

1 2 3 4 5 6 7 8 9 10

About the Book

Since 2000, Industrial Press has been fortunate to include Peter Smid among its family of authors. Both a popular and prolific author, Peter has anchored the CNC portion of our list, headed by the worldwide bestseller, *CNC Programming Handbook*, now in its third edition.

CNC Tips and Techniques: A Reader for Programmers is Peter's fifth book with Industrial Press. During the past decade, in addition to all of his responsibilities as book author, consultant, and educator, Peter wrote a regular column for *Shop Talk*, a monthly magazine that offered "metalworking solutions for the manufacturing and machining industry."

Peter wrote dozens of columns for *Shop Talk*, and his columns addressed the full gamut of CNC topics. They addressed tapping and threading and knurling. They addressed program length and memory needs. They addressed personnel issues, for example, the advantages and disadvantages of having programmers and operators as separate positions or merged. A variety of columns focused on G-codes, M-functions, cycles, and macros. And there were still more.

Ever since *Shop Talk* ceased publication, these columns have been unavailable. Now, in *CNC Tips and Techniques*, Industrial Press is delighted to bring them back to you —over 60 of the best of Peter's columns —- under one cover, so that you can have full and easy access to the advice Peter has been sharing over the past decade.

Each of these essays has been thoroughly reviewed since its initial publication and updated as needed to reflect ongoing changes in the field. They are presented here in chronological order of when they first appeared. An alternative table of contents, organized by topic, has been included, and at the back of the book you'll also find several useful appendices and an extensive index.

We hope you'll find *CNC Tips and Techniques* a valuable addition to your CNC library, and that Peter's advice will make your work more efficient, more effective, and more rewarding.

Peter joins us in welcoming your feedback and questions. Simply write to us at info@industrialpress.com, and we'll be sure to share your messages with him.

Industrial Press
April 2013

About the Author

Peter Smid has been a professional consultant, educator, and speaker, with many years of practical, hands-on experience in industry and education, including CNC and CAD/CAM applications on all levels. His consulting has focused on the practical use of CNC technology, part programming, Autocad®, Mastercam® and other CAD/CAM software, as well as advanced machining, tooling, setup, and many related fields. As a consultant he has successfully assisted several hundred companies with his wide-ranging knowledge.

Over the years, Peter has developed and delivered hundreds of customized educational programs to thousands of instructors and students at colleges and universities across United States, Canada and Europe, as well as to many manufacturing companies and private sector organizations. He has actively participated in many industrial trade shows, conferences, workshops and seminars.

Peter has also written many articles and in-house publications on the subject of CNC and CAD/CAM and for six years he wrote a monthly column in the *Shop Talk* magazine.

Peter Smid is the author of four other best-selling CNC books published by Industrial Press, Inc. and all books are also available as e-books. For more information, go to *www.industrialpress.com*.

CNC PROGRAMMING HANDBOOK
A Comprehensive Guide to Practical CNC Programming
Third Edition – ISBN 978-0-8311-3347-4

CNC CONTROL SETUP for MILLING AND TURNING
Mastering CNC Control Systems
ISBN 978-0-8311-3350-4

FANUC CNC CUSTOM MACROS
Programming Resources for Fanuc Custom Macro B Users
ISBN 978-0-8311-3157-9

CNC PROGRAMMING TECHNIQUES
An Insider's Guide to Effective Methods and Applications
ISBN 978-0-8311-3185-2

Peter welcomes your comments, suggestions and questions. Please write to him at info@industrialpress.com

Acknowledgements

I have made a number of references to several manufacturers and software developers in this book. It is only fair to list them below:

FANUC and CUSTOM MACRO or USER MACRO or MACRO B are registered trademarks of Fanuc Ltd, Japan

Mastercam is a registered trademark of CNC Software Inc.,Tolland, CT, USA

Edgecam is a registered trademark of Pathtrace, Inc., UK

NCPlot is a registered trademark of NCPlot LLC, Muskegon, MI, USA

AutoCad is a registered trademark of Autodesk, Inc., San Rafael, CA, USA

WINDOWS is a registered trademark of Microsoft, Inc., Redmond, WA, USA

There are other companies mentioned in this book.

Caterpillar, Fadal, Haas, Iscar, Makino, Mitsubishi, Okuma, Sandvik, Siemens Yasnac ... and some others that may have escaped me.

Also, my thanks to the publishing team at Industrial Press.
John Carleo, Janet Romano, Robert Weinstein
...without you, this book would not have happened

Table of Contents

About the Book — iii
About the Author — v
Table of Contents by Topic — xiii

2004

CNC Turning — Why Are There So Many Errors?	3
Why Should I Know Manual Programming?	6
Running the First Part — Economically, That Is	9
Are You a CAM Machinist?	12
CAD/CAM or CAD and CAM?	14
Part Program Upgrading and Updating	16
Lathe Cycles — To Use or Not to Use?	19
Short Suggestions for Long Programs	22
Keep Records — Document Your Programs	25
Focus on Numbers	28
CNC Programmer/Operator — Should One Person Be Both?	30
Using CAM Software in Small Shops	33

2005

Minimizing Program Length	43
Conversion of Lathe Cycles	46
Shifting Program Zero — Part 1	49
Shifting Program Zero — Part 2	51
Threading Methods Compared	54
When 1 Thou Equals 65 Dollars	56
Alternate CNC Machine Selection	59
Standard and Rigid Tapping — Part 1	62

2006

Standard and Rigid Tapping — Part 2	67
Mastering M-Functions	70
Tool Length Setup — Three Methods	73
(Extreme) Power of Subprograms	75
Special Purpose G-Codes	78

Well-Structured Program Structure	82
Imaging a Mirror Image	84
Homeward Bound with G28	87
Block Skip Adds Flexibility	91
Simulating the Toolpath	94
Automatic Corner Breaking	97
Working in Planes	100

2007

A Case for Polar Coordinates	107
The "Other" Work Offset	110
Going Helical with Threads	113
G76: Two Formats, One Cycle	116
Multi-Start Threading	119
Automatic Tool Change — ATC	122
Maximum Tool Specifications	125
Control Features — Optional or Standard?	127
Fixed Cycles Repetition	130
Programming Process — When Is It Completed?	133
Quality in CNC Programming	136
Short Ideas and Observations	138

2008

Spindle Speed Control on CNC Lathes	145
Live Tooling on CNC Lathes	148
Trial Cut for Measuring	151
Easing Up on Calculations	154
Preventing Scrap with Offsets	157
Interpreting a CNC Program	160
Default Settings in Macros	164
Create Your Own G-Code	167
Scaling Option	170
Safety and CNC Programming	173
Special Tapping Macro	176
Setting Up a New Part	178

2009

Getting Rid of Chatter	185
Lathe Cycles G70–G72 — Part 1	187
Lathe Cycles G70-G72 — Part 2	191
Limitations in Threading	194
Programming a Long Thread	197
Threading with G76 Cycle — Basics	199
Threading with G76 Cycle — Details	202
Feedrate Adjustment on Arcs	205
Knurling on CNC Lathes	207
Programming a Full Circle	210
Peck Drilling — Watching the Q	213

Appendices

Appendix 1 *Interpreting a CNC Program*	219
Appendix 2 *Default Settings in Macros*	221
Appendix 3 *Create Your Own G-Cod*	224
Appendix 4 *Scaling Option*	226
Appendix 5 *Safety and CNC Programming*	228
Appendix 6 *Lathe Cycles G70-G72 — Part 2*	230

Index 231

Table of Contents by Topic

OPERATIONS

General

Running the First Part — Economically, That Is	9
Preventing Scrap with Offsets	157
Scaling Option	170
Setting Up a New Part	178
Getting Rid of Chatter	185
Peck Drilling — Watching the Q	213

Cutting

Trial Cut for Measuring	151

Knurling

Knurling on CNC Lathes	207

Lathes

Lathe Cycles — To Use or Not to Use?	19
Conversion of Lathe Cycles	46
Spindle Speed Control on CNC Lathes	145
Live Tooling on CNC Lathes	148
Lathe Cycles G70–G72 — Part 1	187
Lathe Cycles G70-G72 — Part 2	191

Machines

Alternate CNC Machine Selection	59

Tapping

Standard and Rigid Tapping — Part 1	62
Standard and Rigid Tapping — Part 2	67

Threading

Threading Methods Compared	54
Going Helical with Threads	113
Multi-Start Threading	119
Limitations in Threading	194

Programming a Long Thread	197
Threading with G76 Cycle — Basics	199
Threading with G76 Cycle — Details	202

Tools

Tool Length Setup — Three Methods	73
Automatic Tool Change — ATC	122
Maximum Tool Specifications	125
Live Tooling on CNC Lathes	148

Turning

CNC Turning — Why Are There So Many Errors?	3

PROGRAMMING

Why Should I Know Manual Programming?	6
CAD/CAM or CAD and CAM?	14
Short Suggestions for Long Programs	22
Keep Records — Document Your Programs	25
Minimizing Program Length	43
Well-Structured Program Structure	81
Control Features — Optional or Standard?	127
Programming Process — When Is It Completed?	133
Quality in CNC Programming	136
Short Ideas and Observations	138
Interpreting a CNC Program	160
Safety and CNC Programming	173

Programming Techniques

Part Program Upgrading and Updating	16
Shifting Program Zero — Part 1	49
Shifting Program Zero — Part 2	51
Imaging a Mirror Image	84
Automatic Corner Breaking	97
The "Other" Work Offset	110
Programming a Long Thread	197
Feedrate Adjustment on Arcs	205
Programming a Full Circle	210

Subprograms
(Extreme) Power of Subprograms ... 75

Toolpaths
Simulating the Toolpath ... 94

CODES, FUNCTIONS, AND MACROS
Codes
Special Purpose G-Codes	78
Homeward Bound with G28	87
G76: Two Formats, One Cycle	116
Create Your Own G-Code	167
Lathe Cycles G70–G72 — Part 1	187
Lathe Cycles G70-G72 — Part 2	191
Threading with G76 Cycle — Basics	199
Threading with G76 Cycle — Details	202

Functions
Mastering M-Functions	70
Block Skip Adds Flexibility	91

Macros
Default Settings in Macros	164
Special Tapping Macro	176

CYCLES
Lathe Cycles — To Use or Not to Use?	19
Conversion of Lathe Cycles	46
Fixed Cycles Repetition	130
Lathe Cycles G70–G72 — Part 1	187
Lathe Cycles G70-G72 — Part 2	191
Threading with G76 Cycle — Basics	199
Threading with G76 Cycle — Details	202

THE MATHEMATICS OF CNC
Focus on Numbers	28
When 1 Thou Equals 65 Dollars	56
Easing Up on Calculations	154
Feedrate Adjustment on Arcs	205

CNC Tips and Techniques

CNC GEOMETRY
  Working in Planes   100
  A Case for Polar Coordinates   107
  Programming a Full Circle   210

WORKING IN THE FIELD
  Are You a CAM Machinist?   12
  CNC Programmer/Operator —
    Should One Person Be Both?   30
  Using CAM Software in Small Shops   33

APPENDICES
Appendix 1 *Interpreting a CNC Program*   219
Appendix 2 *Default Settings in Macros*   221
Appendix 3 *Create Your Own G-Cod*   224
Appendix 4 *Scaling Option*   226
Appendix 5 *Safety and CNC Programming*   228
Appendix 6 *Lathe Cycles G70-G72 — Part 2*   230

INDEX   231

Readings From 2004

A Reader for Programmers

CNC Turning —
Why Are There So Many Errors?

January 2004, updated February 2013

Even in this era of CAD/CAM, many of the part programs that are available for two-axis CNC lathes are still developed manually. There is a good reason why. Purchasing a CAD/CAM system just for a lathe may not be the primary choice of many managers. Modern CNC lathes offer many time-saving features. For example, all centerline operations — such as drilling, reaming, and tapping — are simple point-to-point motions. Lengthy turning and boring cuts can be dramatically shortened by the use of very powerful machining cycles. Special cycles are also available for external and internal threading operations, even for machining castings and for simple grooving.

Yet, with all these programming tools at our disposal, many lathe programs suffer from a number of errors. It is not unusual to see frustrated CNC operators frantically changing programs at the machine, at a great cost to productivity. Why do so many lathe programs suffer this fate?

When I talk to programmers and operators, I get many opinions but seldom real answers. All it takes is a careful look at many of their lathe programs to see that the same type of error appears over and over again. Let's look at them in groups, with some views on what can be done to prevent them:

Calculation Errors

For all their power and simplicity, no lathe cycle will provide calculations of the contour points. When an error is evaluated, typically the programmer either "guessed" or miscalculated. Brushing up on trigonometry is the first step toward improvement in this area. Of all the math subjects, trigonometry is relatively the highest knowledge programmer should possess. I know a few programmers who avoid this obstacle by asking their engineers to give them the contour points from Autocad or some other program.

CNC Tips and Techniques

Other types of calculation errors are accumulated ones caused by incorrect rounding. To avoid this problem, round only the final result — not the intermediate calculations. Many calculations can be confirmed by selecting a different mathematical approach.

Syntax Errors

Syntax errors often occur when the pesky letters O is used instead of the digit 0. These Os are the illegal characters that somehow find their way into the program. For example, the letter Y is not available on a two-axis CNC lathe. Fortunately, the control system will identify these errors.

Logical Errors

Watch very carefully for logical errors because the control system has no means to discover them. For example, a missing decimal point or a negative sign will give a totally new meaning to the programmed dimension. X1 is not the same as X1.0, and Z1.0 is not the same as Z-1.0.

Program Errors

Program errors that are fairly frequent are often those that can cause severe collisions. Making a tool change inside of a hole is one such error. Others include wrong tool or offset selection, excessive spindle speed or feedrate, selection of tool change position, and many others.

Offset Errors

This area is a fertile ground for many frustrations. In my experience, errors relating to the tool nose radius are at the top of this group. If a tool nose radius error is detected by the control system, the cause of this error is always the same — the programmed radius cannot fit into the area provided. This group also represents errors that may be the hardest to identify. Always

A Reader for Programmers

check the offset settings. If the settings are correct, the program itself is at fault. This error often occurs when the tool nose radius is larger than an inside arc of the contour. It also occurs because of insufficient clearances. Because the most common tool nose radii are 1/64 (0.4 mm), 1/32 (0.8 mm) and 3/64 (1.2 mm), always provide a clearance that is at least twice the largest radius. Typically, 0.100 inches or 2.5 mm is sufficient as the minimum clearance. Keep in mind that this is per side, not on diameter.

Setup Errors

Setup errors are strictly the domain of the CNC lathe operator. Check all tools by running the program while the chuck is empty. Are there sufficient clearances? Can the tools index safely? Is the tailstock out of the way? Provide a sufficient grip of the workpiece in the chuck, then watch for the boring bar to exit from a hole safely. Running the first part in a dry run and single block modes may save the day.

Correcting Errors

This area deserves special mention. Many operators fail to identify the actual cause of an error. For example, if a drill does not provide the expected depth, is the programmed depth correct? Is the offset set correctly? Is it even possible that the drill was pushed slightly into the holder? Identifying the real cause of a problem is the first step towards its elimination.

I hope by reviewing some of the common errors found in many programs for CNC lathes that you can look at your own programs and see them in a much different light.

CNC Tips and Techniques

 ## Why Should I Know Manual Programming?
February 2004, updated February 2013

In older days, most CNC programs were developed manually, using simple common and everyday aids — pocket calculator, pencil, paper, and the proverbial five-pound eraser. Combine this human effort with punched tapes and some extra hardware, and you have what seems today a history long past. The tape is gone today, and even though many programmers still sharpen their pencils and purchase new erasers, the times have clearly changed in favor of computer-based programming, or CAM programming. True, CNC programming using personal computers started in the mid-1980s, but initial costs, the state of computer technology, long learning curves, and a lack of features prevented their wide usage for some time.

Using CNC software to make part programs is not a novelty. Feature rich software from a number of vendors offer the whole gamut of options — from entry level 2D to sophisticated 3D, solids, multi-axis machining, and multi-tasking — all available for desktop and laptop computers. Public and private institutions offer many training programs in this area, often in addition to standard CNC programming courses.

This brings up the ever more important question that many students ask: *"Why should I know manual programming, when I am learning Mastercam?"* Of course, there is software available other than Mastercam, but the question raises a good point. I wish I had a dollar for every time this question pops up in conversation or during a training session. So what gives?

Every CNC programmer has to possess certain skills. One skill I consider the most important is the *ability to machine a part*. Deciding on the best setup, selecting the most suitable tools, and creating safe and efficient machining operations is *the* core skill of any successful CNC programmer. The actual method of programming has nothing to do with it. Computers cannot make intelligent decisions.

A Reader for Programmers

Although the part program, in the form of a file or printed copy, is designed for a particular CNC system, we — the programmers and machine operators — have to be able to understand it.

Let me go through five main reasons for this statement:

Reason 1 — Myths about Post-Processors

Most CAM software comes with pre-configured post-processors in the form of standard text files. These files are *intended* to be customized by the end user. There is no way for any software vendor to cover all possible combinations that exist between machine tools and control systems. In addition, vendors cannot predict the individual preferences of each user. As a post-processor is used to format the final output of the part program, you still have to *know* what that output should be.

Reason 2 — At the Control with No Control

When the program runs at the CNC machine, you can watch it scroll by on the control display screen. Do you understand what all the letters and numbers mean? What do you do when there is an error in the program? CNC operators who can make minor (and sometimes even major) changes to the program are much valued by their employers. Some operators do not mind just pushing the buttons, but if you want a certain level of control and responsibility, you have to be able to understand the program itself.

Reason 3 — Programmer / Operator as an Opportunity

Small shops and job shops often do not have the luxury of employing a full time CNC programmer. As important as such a person would be to the company, many owners and managers opt for a qualified CNC operator who can also prepare the part programs, then run them. This programmer/operator concept has many benefits for both the employer and the employee. Combine Reasons 1 and 2, and you have a Reason 3.

CNC Tips and Techniques

Reason 4 — Do You MDI?

Any operator who sets up a CNC machine has to perform many pre-machining activities, for example, make tool changes, set tools, activate the spindle, orient it, move the table by a certain distance, check a position, and so on. Even with many switches on the panel, one of the best methods is to use the *Manual Data Input* feature of the control — the MDI. This feature allows you to input and execute data at the control on a temporary basis. The data is always provided in the program format of that control. You cannot use the *MDI* feature if you do not understand at least the basics of manual programming.

Reason 5 — Learning from others

A part program printed on paper does not teach programming or machining methods, yet there is so much to learn from programs that have already been verified. How many of us started our careers by adapting existing programs to our own? A part program that works well offers tremendous opportunity for study and learning. Of course, one has to be able to read such programs — *to interpret them* — and that means being able to understand the G-codes and M-functions, the program format and structure, offsets, modes, and the many addresses and their subtle meanings.

Of course, one has to be *willing* to learn. When you just look at a part program, without understanding it, without being able to interpret it, do you care? Does it bother you? Do you feel you want to know more? If the answer is "yes", you are on the way of becoming a better programmer or operator — a true professional. But that is reason number six, and I only promised five.

A Reader for Programmers

Running the First Part — Economically, That Is

March 2004, updated February 2013

Machining a batch of the same parts in a single production run is probably the most common reason for purchasing a CNC machine. The CNC technology offers a great amount of predictability. There is consistency between parts, dimensional accuracy is maintained over many parts, and the program and setup — once verified — can be used over and over in the future.

Yet, even with proven part programs, the production always hits one weak spot when running a new batch. Yes, it is the *first part*. Every supervisor knows that the first part of the batch is also the *most expensive part* of the whole batch; its cost of manufacturing proportionally affects all remaining parts. To minimize the cost of making the first part, it is worth looking at several influencing factors and possible solutions.

Comparing Programs

There is a significant difference between running the first part using a new program and a program that had already been verified.

New part programs

CNC programs that have never been used must be carefully inspected. Even programs generated by a CAD/CAM system require certain scrutiny, although on a much smaller scale. Programming errors and oversights can find their way into any program in various forms, some quite hard to detect. For example, the depth and width of a cut may be out of a reasonable range, clearances may be too large or too small, spindle speeds and feed rates may be overrated or underrated, a tool may not be the most suitable one for the job, and so on. These oversights are virtually impossible to see in print or even on the screen — they

CNC Tips and Techniques

show only when the first part is run on the CNC machine. In addition to the program itself, the machine setup is always new, regardless of whether the program is new or previously verified.

Verified part programs

These are part programs accepted for production and verified at some earlier date. Keep in mind that only the programs have been verified, not the setups or tools. Setups and tools are subjects to frequent changes; they must be verified every time. Another important consideration is the CNC machine tool being used. Even verified programs have to be checked if they are used on a different machine than the previous time.

Decreasing Costs

Decreasing the cost of the first part — and, therefore, the whole batch — has to be considered at four levels, all interrelated:

Planning

Process planners or supervisors have to evaluate many aspects of the production. One of their key considerations is the number of parts produced in a single batch. The more batches per period of time, the more expensive each batch will be, mainly because of the repeated setups. Assuming that more parts produced in a batch will have no negative side effects (for example, on inventory), fewer batches per period of time might provide part of the solution.

Programming

CNC programmers can literally do wonders to decrease the cost of the first part through the program. Here are only some ideas. Using consistent tool numbers for commonly used tools saves time during setup, as tool registration and many related offsets will not change. Another very effective programming method is to include one or more trial cuts with the block skip function in the program. These cuts will only be used for the first

A Reader for Programmers

part testing, but turned off for a full run (except the occasional inspection). Don't forget to provide sufficient comments in the program, so the operator knows the purpose of these special cuts. Another consideration, especially for programs developed manually, is to invest in reliable CNC simulation software. Although no software can simulate every detail of CNC machining, it can be an excellent tool to find errors and areas of improvement.

Setup

At the machine, the CNC operator is normally responsible for setting up tools and fixtures, registering tool numbers and various offsets, troubleshooting the program, and monitoring the operation. Speeding up the non-productive time required for setup is the most significant single saving available. Modular fixturing, common tooling, general setup consistency, and the operator's skill all add up to make the transition between setup and operation faster and smoother.

Machining

During machining, the CNC operator relies heavily on the program itself. During the first part run, it should not be enough for the CNC operator to look just for errors or monitor the program flow. Operators can and should do much more at the same time. For example, they can think about optimizing the program for better performance. Speeds and feeds are the most common areas to focus on, but many more "little things" also influence the cost of manufacturing (including the first part).

Lowering the cost of the first part should be a team effort. CNC programmers may be the key people in such efforts. However, mutual cooperation of everybody involved will offer many benefits and have a positive effect on the proverbial bottom line.

CNC Tips and Techniques

 ## Are You a CAM Machinist?
April 2004, updated February 2013

The main topic at a discussion I had several years ago was quite simple: *CAD/CAM Learning Curve* for CNC part programming. How long does it take to learn *CAD/CAM* software to produce a quality part program for a CNC machine tool? This type of part programming is often associated with the term *CAM programming*. Having looked at many courses offered by many more providers, the picture is still a bit cloudy. From an introductory three-day course (eighteen hours in reality, breaks included) to a three-level course running more than one hundred hours, there is something for everyone. All that in addition to self-starters, books, and multi-media resources.

Learning a CAM system can be quite intuitive for someone experienced with computers. For such a person, creating geometry (mainly points, lines, and arcs) can be child's play after a few short hours. On the other hand, the same CAM software can be somewhat intimidating for somebody whose experience is with machining parts but not with using computers. That brings up an important question: *What is required from a CAM programmer?* What type of knowledge and skill should a CAM programmer have in order to develop often complex tool paths that make parts on expensive CNC machines?

During the discussion I mentioned above, the word *skills was* indeed used many times, mostly related to computer skills. Even the term *creating geometry* (also used quite frequently) is simplistic. The correct phrase should be *creating toolpath geometry* — and here lies the big difference. It is fairly easy to develop a geometry that looks right on the screen. It is not easy to create a toolpath geometry that guides the cutting tool and can often make or break a small shop.

In CAM programming, the purpose of geometry is not to re-create a paper drawing as many beginners try to do. Its purpose is to create a *working toolpath*. True, most CAM systems have hundreds of features that make the entire process simpler and

A Reader for Programmers

faster. Features such as lead-in and lead-out, multiple contouring passes, segmentation of depth cuts, and stock allowances all contribute to a software that becomes easier and more efficient to use. Their inclusion requires another important skill — understanding the actual purpose of all these features. You must understand when to use them and how to use them correctly. You must also have a solid knowledge of machining techniques.

There are many times when even the best features of CAM software may not provide the desired result. Often it is necessary to add a bit of geometry here and there, a little extra tweak to get the exact result required. The part programmer should have an exact vision of the tool path in mind — before even starting to use the computer.

With all this in mind, what is a reasonable learning curve for using CAM software? Is the computer guru in the shop, who has minimal if any knowledge of machining, better suited for the task than the experienced machinist with limited computer knowledge? Most people will agree that learning how to operate and use a computer and software may take a few weeks at the most, whereas learning the art of machining takes years, even a lifetime.

The learning curve for any individual will also depend on other personal and professional skills. Even in CAM programming, there is always some math involved. There is also the entire area of understanding productivity, efficiency, and safety issues in manufacturing. On the technical side, knowledge of the machine tool, the control system, the fixturing, tooling, setup methods, speeds and feeds, and so on, is most important for each and every program developed. In a nutshell, the same qualities and skills required from a part programmer using manual methods are required from a CAM programmer. Unfortunately, this is not always the case.

So who is this CAM machinist from the headline? I believe that the inescapable truth is that you have to know *how to machine a part* before you can be a successful CAM programmer.

CNC Tips and Techniques

You have to become a part programmer with high machining skills — a true CAM machinist.

 ### CAD/CAM or CAD and CAM?
May 2004, updated February 2013

In various forms and on various platforms, CAD/CAM is not a newborn idea. Even before Personal Computers (PCs) started occupying every designer's desk, Computer Aided Design/Computer Aided Manufacturing software — known simply as CAD/CAM – was available on expensive mainframes and mini computers. Needless to say, it had a matching price tag. This was the dawn of a new way of design. The convergence of design and manufacturing became a hot topic. Designing a part in CAD and developing the tool path in CAM seemed like the way to go for the foreseeable future. Most Fortune 500 companies applied CAD/CAM to manufacturing and it worked very well for them in the competitive market place.

What is the situation today in small and medium machine shops? How do they handle the design and the toolpath convergence? Before I get into it, you have to consider one major advantage that the large manufacturing companies have — *they have their own product.* They have their own in-house team of engineers, and they generally use a big CAD/CAM system. The success of such convergence is measured only by the effectiveness of each company.

Overall, small shops are of two kinds, those that develop their own products and those that do not. The latter ones are often called *job shops* or *custom machine shops.* The shops that have their own design capabilities should have no problem with the CAD/CAM convergence, if they manage it right. Many job shops take advantage of a computerized toolpath generation, but have no real need for CAD software. Still, they often face a string of obstacles to achieve the same goal.

A Reader for Programmers

Consider the realities. Your customer faxes you a drawing, asking for a quote. Your quote wins and you get the work. Often that faxed drawing is the only drawing you have from which to make the part. In a better scenario, the customer sends you a clean paper drawing. Or better yet, the customer sends you the drawing in the form of a disk file. Now, that is the best scenario, but — yes there are "buts" here.

As most work in the job shops is two dimensional (2D), the design is often done in CAD software such as Autocad by the customer. Shops that have CAM capabilities use a tool path development software such as Mastercam, EdgeCam, or GibbsCam. The question is how to get the drawing file into the CAM software. The ideal method is that the software reads the file in its native format. For example, the popular Mastercam can read and write Autocad DWG files directly, without a translator. If the software cannot read the native format of the original, it has to have a reliable translator.

Virtually all systems support DXF (drawing exchange file) and/or IGES (Initial Graphics Exchange Specification). DXF is quite simple and suitable only for two dimensional drawings that are composed of lines, arcs, circles, and points. IGES is more powerful and used for translation of complex three-dimensional drawings.

So what is the best approach for a small shop to take?

Consult with Your Customer

Consulting with your customer is probably the most important step. Your customer likes to know about your capabilities. In turn, you should know what the customer can offer to your shop, particularly when it comes to a drawing supplied on a disk or by email. Ask for a file format that your CAM software supports, preferably in the native format.

Understand CAD/CAM Translators

If the translator has to be used, make sure you understand

CNC Tips and Techniques

its capabilities and its limitations. Poor translators may not convert the original geometry, especially when it contains complex geometry such as splines.

Train your CNC Programmer

Good CNC or CAD/CAM programmers can detect drawing flaws and mistakes, and often solve any problems quickly. They will also be able to eliminate geometry that is part of the drawing but not necessary to create the tool path. This skill might be one of the programmer's most important skills.

Educate Your Customer

Do not be afraid to tell the customers what you expect from them. They will understand that your requests are made in the spirit of producing their part within specifications as well as efficiently.

Yes, it seems that for many reasons true CAD/CAM convergence is not a part of small job shops. At the same time, there is nothing wrong with CAD and CAM working together well in this unique environment.

Part Program Upgrading and Updating
June 2004, updated February 2013

Regardless of the methods used, developing a CNC part program does take time. Whether written manually or generated with the aid of a CAD/CAM system, a part program should not be considered as completed until it is used to run a few parts and optimized. Even the best part programmers cannot always predict every condition during actual machining. It is not unusual — in fact, it is very common — to see CNC operators make changes to the program at the control. If the program is perfect, no changes would be necessary, a situation that rarely happens.

A Reader for Programmers

Optimizing a part program means improving it, mainly for more efficient performance, but also for other reasons, such as change in setup, use of a different tool, or even improved safety. Optimizing a program can take place at the control (usually by the CNC operator) or away from the control (usually by the CNC programmer). Two terms are often associated with a program change — *program upgrading and program updating.*

Program Upgrading

Program upgrading means strengthening the program, enriching it, and making it more cost effective without compromising quality of the part or safety of machining. When running several thousand parts in a batch or a job that repeats from time to time, upgrading part programs will have a profound effect on the overall cost of doing business. Shortening the cycle time by a few seconds can mean hours in overall savings. Here are some ideas that programmers and/or CNC operators can consider when upgrading an existing part program.

Programmed spindle speeds and cutting feedrates are the first items to evaluate; they require a very small intervention with virtually immediate improvement. It is a well-known fact that programmers take a rather conservative approach in this area. Another area of interest should be the clearances applied in the program — "cutting air" is never productive and should be minimized. Changing a grade of the cutting insert and increasing the feedrates also offer benefits without making major changes to the program. If the program contains various machining cycles, operators can make even more significant changes, depending on the cycle type.

One of the overall cycle time killers is excessive dwell time. The minimum dwell time is the amount of time required to complete one spindle revolution. In practice, this amount is often doubled to allow full revolution dwell at 50% of programmed spindle speed. For example, a programmed dwell time of one second for 1200 rpm spindle speed is excessive, yet often programmed. In this case, the minimum dwell required is 0.05 of a second, or 0.1 of a second in practice. For a few thousand parts, the time saving will be considerable.

Rapid motions can also shorten the cycle time. Combining two single motions into one simultaneous motion will also make the cycle time shorter; so too will making a tool change without moving to machine zero in all axes. There are other areas of upgrading that can be explored, some specific to a particular machine tool, for example, using a fewer threading passes on a CNC lathe.

Not all program upgrades can be made with the same ease as the suggestions provided here. For example, increasing the depth of cut may require a rewrite of the program code for a specific tool. Some changes can be possible only if the control system offers certain optional features. Equally important is knowing when not to upgrade. Program upgrading for a small non-repetitive batch of parts may be counterproductive and not worth the time required to make the changes.

Program Updating

Program updating is not as common as program upgrading. It involves a change to the program due to a change in the engineering drawing. This situation occurs more frequently in companies that design and manufacture their own product. Jobs shops also experience a drawing change, often initiated by the customer or after mutual consultation with the customer. Often a single dimensional change may require a major program update. Talk to the part designer or engineer about tolerances that appear too tight. It is not unusual to find tolerances that can be opened up — we all know that the closer the tolerance, the more expensive it is to make the part. With more significant changes to the drawing, a complete program rewrite may often be required and worth the extra time.

The last, but not least, consideration is the quality and skills of the CNC programmers and the CNC machine operators.

A Reader for Programmers

Good training combined with experience will reward every shop owner in a relatively short time. As a shop owner, try to provide your employees with quality training — it will pay off in no time.

 Lathe Cycles — To Use or Not to Use?
July 2004, updated February 2013

If we have to name a single major benefit that modern CNC technology has offered over the initial NC technology, it would have to be the various forms of machining cycles. Yes, those small routines save much of programmers' time and calculations. CNC machining centers have had their ubiquitous G81, G82, G83, etc., for drilling, boring, tapping, and other operations related to hole machining for a long time. This group of special cycles makes perfect sense for the part programmer because even companies specializing in complex 3D can use them efficiently, albeit occasionally. Virtually every CAM software program supports these *fixed cycles* or *canned cycles*, as they are commonly called, through the post processor output. The story is somewhat different when it comes to CNC lathes and turning centers.

Multiple Repetitive Cycles

In the early 1980s, Japanese company Fanuc adopted a concept long known to part programmers from the 1970s, who were using various programming languages, such the venerable Compact II, to develop roughing and finishing tool paths. The mathematical concept behind this development was quite simple — define the stock boundary, define the boundary of the finished contour, and the control will remove all the material contained within these two boundaries. The result of this adaptation of an early technology into a modern CNC turning system is now known under the collective umbrella of several *multiple*

CNC Tips and Techniques

repetitive cycles.

Without a doubt, three of these cycles have become the staple of manual programming for CNC lathes since day one: the G71 cycle for roughing, the G70 cycle for finishing, and the G76 cycle for threading. Their ease of use (the user calculates only the finish contour points) and the possibility of operator changing the cutting parameters at the machine control have made them instant winners. The strongest feature of these cycles, brevity (only one or two blocks of program are required for roughing), also became their small disadvantage.

CNC lathe manufacturers and vendors have realized that they can offer a better price on their machines if they make the available memory smaller (CNC memory is very expensive even today). There are still many CNC lathes that have a very small memory storage capacity for the part programs, although that is changing rapidly.

Yet, even if the issue of capacity is becoming a non-issue, there is always the benefit of being able to change various cutting parameters at the control (depth of cut, for example). The developers of CAM software have realized this need and generally support the output of multiple repetitive cycles in the final program. What many developers also offer is the option of not choosing the cycles as the program output. Selection of this option will, of course, require a more robust memory capacity. It will also bring up an important question regarding *when not to use these cycles.*

Cycle Programming

Cycles are nothing more than pre-programmed routines. They will define the tool motions strictly based on the criteria included in these internal routines. To be blunt (and somewhat unfair), the programmer is at the mercy of the software engineers who developed the control unit. Make no mistake about it. The control software was developed for most — but not all — turning, boring, or threading applications; it does the job extremely

A Reader for Programmers

well in such situations. Logically, the software *could not* be developed for special applications.

Yet, programming is somewhat like sculpting. There is technology involved, there is skill involved, and there is strong individualism involved (called art in sculpting). In order to control every single tool motion exactly as planned, the programmer may find a particular cycle to be ineffective.

Take, for example, the G71 roughing cycle. It starts at the beginning of the cut and continues towards the end, with the change of direction allowed along the X-axis only. If your program requires frequent changes of cutting direction, the longer method (block-by-block) will be the choice. Single point threading cycle G76 is no different. It is great for a single start standard thread, but loses its appeal somewhat for multi start threads (although it can be still be used). It becomes virtually useless for special threads such as those with special forms, particularly when the threading bit is smaller than the thread itself or the thread shape has to be interpolated rather than formed.

Without a doubt, these multiple repetitive cycles have made the programming and machining jobs much easier for all of us. At the same time, we have to understand their limitations and the alternatives for situations when these cycles are not preferable or are even impossible to use efficiently. Apart from programming every tool motion manually (which is always an alternative), make sure your CAM system can handle the type of tool motions required by the work you do and your programmer can generate the critical tool path. Also make sure you have enough memory available in the control system to store these programs — they can be quite long. A DNC (direct numerical control) solution may help as well.

CNC Tips and Techniques

 ## Short Suggestions for Long Programs
August 2004, updated February 2013

Once in a while, your machine shop will face the need to run a CNC part program that exceeds the built-in control memory capacity. This method is commonly called DNC (direct numerical control) and uses the TAPE or EXTernal mode of the control panel rather than the more common MEMORY mode. Not much is required for this method to work with your CNC machine — a suitable software, a desktop or a laptop computer, a connecting cable and some configurations. If all goes well, you are up and running in no time; if not, read on for some suggestions.

For the purpose of this column, I define the term *long program* as a part program that cannot be executed from the control memory. The most common source of such programs is a CAD/CAM system processing complex machining operations. Although the majority of long programs involve some 3D machining (for example, mold work and multi-axis toolpaths), there is a growing need for more control of 2D toolpaths, for example, in pocketing or at high speeds. High quality CAD/CAM software offers a multitude of toolpaths and pocketing styles. When combined with control of the cutting width and depth, size of the part, cleanup of corners, ramping, etc., it is not unusual to end up with a long program. The main difficulty when running such a program is when the CNC operator has to *restart* the cut somewhere in the middle of the program, after a substantial tool path had been completed with a particular tool. Often, the main reason is a tool breakage.

Most CNC machines do not have *program restart* or *sequence return* features which address this problem. Yet, there are a number of ways the CAD/CAM programmers can minimize the downtime associated with interruption of the long program. Let's look at some possibilities to consider:

A Reader for Programmers

Computer Location

Locate the computer that sends the long program as close as possible to the CNC machine that receives it. As a rule, the computer that receives the data (CNC) must be ready *before* the computer that sends the data (desktop or laptop). That means a bit of running between the two that can be eliminated.

Program Sections

This topics is a very important consideration. Each tool used in the program is automatically a separate section. However, sections for a single tool, especially a tool that does a lot of work, should be carefully defined as well. One technique that can be used is homing the machine (machine zero return) after each section has been completed, particularly along the Z-axis, then starting at the next section with the same tool. As these extra motions add to the cycle time, they should be preceded with a slash (block skip function). An addition of program stop (M00) or optional program stop (M01) to this method may offer additional benefits.

Program Editor

Use a dedicated CNC editor (designed for CNC files) or a text editor that can highlight individual addresses in the program — for example, all Z-axis motions in red. Colors help to identify individual sections, and the editor can also be used to create a temporary copy of a shortened program.

Block or Sequence Numbers

They do help a lot. Although blocks or sequence numbers add to the overall length of the program, this length is irrelevant in DNC operations. If a tool breaks, the sequence number at the time of interruption offers some information that can be used for restart. Instead of using block numbers for each block, consider using only a few block numbers in strategic places within the

CNC Tips and Techniques

program. A good place for selective block numbers is at the beginning of every new section that uses the same tool, or at home position.

Descriptive Comments

Identify individual operations or sections of the program. For example, if the same tool is used for roughing five pockets, place a comment at the beginning of each pocket tool path with a unique description.

Feed and Retract Planes

Be consistent in selecting the Z-axis positions for the start of cut and retract when the cut is completed. Text editors can be used to search and/or replace common data.

Cutting Feedrates

If you make the plunging feedrate different from the cutting federate, you will more easily find the beginning of a section from which the change can be done, just by searching for the feedrate. Also, you will not accidentally forget a feedrate or use the wrong one.

What else is there to watch for? The above suggestions can be applied individually, or combined. The most important is to make sure that any change implemented is a safe change. Watch for *XY* tool locations, missing spindle speeds and feedrates, the status of switches on the control panel, offset settings, and other related subjects.

Running a long program from an external computer in DNC mode can be very simple and uneventful. On the other hand, if problems do arise, be prepared to deal with them via good planning in order to minimize the undesired downtime.

A Reader for Programmers

Keep Records — Document Your Programs
September 2004, updated February 2013

There are many reasons to keep track of work done during both common and not-very-common manufacturing processes and activities. The most important reason to keep track of the multitude of activities is to create some reasonable documentation for the work in progress. The method and its purpose are quite basic — to be able to register and retrace or retrieve individual development steps or stages in case of future need. Whether used for emergency purposes or general reference, good documentation is imperative for any serious work in a machine shop environment. At the corporate level, documentation is often developed for internal purposes only. Documentation can also become a major part of the ISO 9000+ certification process, in which case it takes its own form. At the level of CNC or CAD/CAM part programming, the needs in this field are much more relaxed and flexible, but equally important. This essay looks at program documentation relating to CNC programming, regardless of its method of development. Good documentation is as important for manual programming as it is for CAD/CAM tool path development.

Most part programs reach the CNC operator in one of two forms, as a printed hardcopy or as a disc file. In both cases, the program format is the same, only the method of distribution varies. All CNC systems offer a method of inserting comments and messages into the body of the part program. If used properly, these comments and messages are located at strategic locations within the program, as comments intended for the CNC operator. The most common way to include comments and messages in a CNC program is the use of matched parentheses (). This (COMMENT) method is typical for Fanuc and related controls. For example, the following message instructs the CNC

CNC Tips and Techniques

operator to check the width of the part:

(WIDTH MUST BE BETWEEN 26.7 AND 27 MM)

Other methods use the dollar sign ($) or the semicolon (;) and similar methods with the same purpose. In all cases, the control system is designed to ignore the message, or comment, during program execution.

Having too many or too few comments in the part program can be counterproductive. Having no comments at all in the program can potentially lead to wrong assumptions, unless some other way of information is provided. The main goal of CNC program comments and messages is to provide an information link between the CNC programmer and the CNC operator. Thorough and informative program documentation is so important that every shop owner or manager should insist upon it and every programmer should provide it as a standard program feature.

Good Program Documentation

What are the main characteristics of good program documentation? The first rule is to make any comment in the program short and to the point. The second rule is to include only those comments that are relevant to the project on hand. In terms of contents or subject matter, here are some suggestions that many CNC programmers have used successfully in various program comments:

Date and Time

The current program version is critical; it must match the part drawing. The latest date and time are assumed to be the latest version.

Programmer' Name

The programmer's name is important to the CNC operator in case questions or clarifications are necessary.

A Reader for Programmers

Drawing Number and/or Revision
Include the latest drawing version used during the programming process.

Machine Type
A well-developed part program can be used on CNC machines that are similar. A program comment should list which machine/control combinations will accept the given program.

Setup Data
Specify the part zero in all axes. Include the setup method and all related information, such as part orientation, depths, and clearances.

Tooling Data
List all tools at the program beginning and repeat the tool description for each tool in the program.

Special Instructions
If the program contains M00 (program stop), describe the reason for its existence. For example, instruct the CNC operator that the part is to be reversed, or special lubrication has to be applied for tapping, or that the tool has to be checked. Many similar examples fall into this category.

There are many different applications that benefit from comments and messages inserted in the part program. Depending on the control system, the comments may not load into the control memory, but they can always be available in printed form or in the program file. CNC programs developed by CAD/CAM software should include the same comments and messages as programs developed manually. All high level CAD/CAM software offers the insertion of comments into the program.

A well-documented program is easier to interpret by the CNC operator and even by its own author (the programmer). Documentation should be provided as a standard feature of each program.

CNC Tips and Techniques

Focus on Numbers
October 2004, updated February 2013

There is a good chance that the drawing from which you get to make a part has originated in CAD software. Nothing fancy, just a standard, everyday-type 2D drawing — two, three views, dimensions, and some text — that's all. However, just because the drawing has been prepared in CAD does not mean it is without flaws and even errors. Let's look at this fictional drawing and focus on a few details relating to numbers (dimensions, actually).

Focus 1 — Metric or Inch?

What units are used in the drawing? If you have no clue, don't assume — make sure! More drawings are now prepared in metric units than before, but they are not always prepared properly. The magic word METRIC somewhere in the corner of the drawing may just not be there to tell you. Here are some ways to tell (but verify it later, anyway).

The first clue is the drawing scale and the second clue is the dimensions, their nominal values. If your specialty is producing small parts and you see the part length as 50, chances are you have a metric drawing. If you are used to seeing dimensions such as 1.625 or 0.75 (inches), and there are no dimensions that small, chances are you have a metric drawing. If the scale is correct, it solves another piece of the puzzle. Another clue is the number of decimals used in dimensions. If the drawing follows proper standards, a metric drawing should have whole numbers without the decimal point. So, our length of 50 mm will be shown as 50 between the dimension arrows, not 50.0. This is how it should be, but not always how it is.

Which brings us to yet another clue, and that is how decimal numbers are expressed. Because the metric system uses 1/1000th of a millimeter as the smallest unit, you should never see more than three decimal places in a metric drawing — and even those are rather rare. Sometimes you have to convert

A Reader for Programmers

dimensions if most of them are in one unit, but another (a thread or tap size, for example) is in the opposite unit. It can be messy out there, but that's the price we pay for being behind the rest of the world in adopting the metric system.

Focus 2 — Tolerances

A related item that could cause trouble — tolerances — is also directly related to a CAD system. Let's stay with inches for this example. The drawing you have just received is basically a rectangular part, with overall dimensions shown as 5.0000 x 3.0000 x 0.5000. What is this? A few decades ago we would have been taught that the number of decimals places indicates the precision of tolerance. For example, x.xxx would be ±0.001 (or whatever the internal company standards were then). So now we have a part that has four decimal places on all three overall dimensions! That's not all — there is also a surface finish of 125–250 required on the length and width. Because this dimension makes no sense, we have to look at the real culprit behind this.

First, throw out the old notion of this tolerance being implied by the number of decimal places. Yes, it is still done and, yes, it is still wrong. Separate from the fact that the notation breaks all ANSI or ISO standards, it is wrong because it is not reliable. It is not reliable for at least one reason (there are several more) — because a CAD software defaults to four decimal places for inch drawings, and to three decimal places for metric drawings. This can develop some bad habits in the CAD/CAM community. Of course, the software does allow for changes to any number of decimal places, but it is the role of the CAD operator to choose the proper settings.

Focus 3 — Why Is Metric Better than Inch?

Note the question again. It does not ask *if* metric system is better, it assumes that *it is* better, and asks *why* it is better. Before I get too many e-mails challenging my question, let me state for the record that I am specifically referring to CNC machines, not

CNC Tips and Techniques

to any personal habits of buying a one-pound T-bone steak at your supermarket (although I could argue that one, too). CNC machines can be set to either inch mode or metric mode through the part program or at the machine control itself.

What is important here is to understand something called the *minimum increment* of the control system. In plain language, it means the amount of the smallest motion possible within each unit selected. For typical CNC machining or turning centers, the amount is set to 0.0001 of an inch or 0.001 mm, known as one micron. Now, say you want to make an offset adjustment at the control of 0.00004 of an inch. You just can't unless you program the part in metric. The amount of 0.00004 of an inch is equivalent to 0.001016 mm, which is a motion possible to make, as one micron. The machine will move by the equivalent of 0.00003937 inch, close enough to pass all inspections. Designers out there — take a note!

CNC Programmer/Operator — Should One Person Be Both?
November 2004, updated February 2013

When it comes to jobs in the CNC field, two types of positions within a plant or machine shop are usually identified: *CNC Programmer* and *CNC Operator*. Many small or even medium-sized machine shops do not have the resources or even the need to keep each position separate. Instead, they combine them both into a new position, commonly known as the *CNC Programmer/Operator*. The first apparent and most obvious benefit is saving money on wages or salaries. However, there are additional benefits and the inevitable disadvantages that are worth considering before creating such a position.

A Reader for Programmers

Advantages of CNC Programmer/Operators

First, let's look at the advantages. Involving one person in the larger process of making a part provides better understanding of that process. If problems do occur, solutions are often easier to find. Programmers who also run the CNC machine are more motivated to do a better job using their own program. It is a common requirement for part programmers to know how to machine a part, but it is not normally required of CNC operators to know all details about how to develop the program. Good CNC operators do not always understand the programming part and cannot always offer program-related solutions.

In a combined position, this lack of understanding is eliminated. Programmers also frequently interact with engineers and customers; operators usually do not. Good programmers also interact with machine operators and seek their ideas. Again, the combined position may offer solutions to design or manufacturing process that only skilled machine operators could think of.

Disadvantages of CNC Programmer/Operators

There are also disadvantages to this programmer/operator concept. The initial appeal of saving money on remuneration may be offset in the loss of production time. It is virtually impossible to make a program and at the same time operate a machine. To do both equally well takes not only a skilled and dedicated person, it also requires a suitable job. If a part has to be changed on the machine every two minutes or so, the programmer cannot sufficiently concentrate on the work of developing the next program. A part that takes twenty minutes to machine offers a little more flexibility.

Every shop manager knows the term *machine utilization* — that the road to profitable business is running CNC machines as much as possible, producing quality parts. There are always some delays involved when using CNC machines. First, it is the setup time, including the fixture, tooling, and control settings. Next, the first part has to be verified. Then when the machine

CNC Tips and Techniques

does finally start production, there is always the necessity of changing parts, replacing dull inserts, modifying offset values, etc. Adding more idle time because the programmer has to prepare the program can be very costly; over a period of time it can cost more than having two skilled workers taking on individual responsibilities.

Thus far, this overview has applied to a single CNC machine. What if there are two, three, or even more machines in the shop?

Three other disadvantages of the combined position should also be considered very seriously. The first is quite simple. If the programmer stays home sick, so does the operator (and vice versa). The second disadvantage is related. What if the programmer/operator quits? Of course a suitable backup is an ideal solution, but the reality is that there will be a gap in the production somewhere; even if the gap is temporary, it still represents time lost. The third disadvantage relates more to certain traits in human nature and does not apply to everybody. Some programmers who also run the machine may have the tendency to cover up their mistakes, spending additional time to make fixes at the machine, even a new program, without anybody's knowledge. Some programmers/operators may feel self-satisfied or even complacent; they stop learning new methods and techniques. Others may have exaggerated demands in terms of wages, benefits, time off, and so on.

One item that needs consideration is the method by which the part program is generated. It is still common in many shops to write the program manually rather than using a CAD/CAM programming system. Although is it important for the programmer to understand manual programming, it is equally important for the shop manager to understand the benefits of a CAD/CAM (also known as CAM) programming system. For the combined position of one person programming and operating, a CAM system installed in the vicinity of the CNC machine may bring many benefits, in spite of its initial cost.

A Reader for Programmers

A number of machine shop managers prefer the CNC Programmer/Operator position and are well aware of its drawbacks. If established in a suitable environment and managed properly, having a multi-skilled worker in the shop should bring in more benefits than problems. The majority of CNC programmers who also operate the CNC machine are honest and dedicated; they do their best to do the job well. They consider themselves true professionals, and rightly so. On the management side, a prudent owner or manager has to consider all pluses and minuses before creating the position of a CNC Programmer/Operator.

 ## Using CAM Software in Small Shops
November 2004, updated February 2013

Over the years, manual programming has offered many benefits to CNC programmers. It offers absolute control over the toolpath and it teaches lessons in organization and discipline. Also, it does have some negative side effects; it is very time consuming, prone to human errors, and hard to make changes in the program. The need to eliminate — or at least minimize — the problems associated with manual programming led to the development of CNC programming using computers, known as *CAM programming*.

CAM SOFTWARE

Modern CNC software — typically called CAM software because of the many trade names ending with the three letters CAM, such as the popular Mastercam — offers many features that automate the original manual method of programming. Computers produce most CNC programs more quickly and more accurately than manual programming ever did. A proper application of CAM software will produce a part program that reflects the programmer's way of machining a part.

Whether developed manually or generated by software, the

CNC Tips and Techniques

part program has to be in a format compatible with the CNC machine control unit. Comparing manual programming with CAM programming is not fair; however, the knowledge and understanding of manual programming techniques is important in all CAM applications.

Small shops are not like big companies. Their needs are different and so is their management style. Desktop or floor-based CAM programming is not new — it has proven itself in many machine shops. Software development has been steadily improving. Many simple and even complex programming applications are available for much lower cost than only a few years ago.

What Is Involved?

Hardware is the physical parts of CAM. However, hardware is not the focus of this article — it changes so rapidly that something new will be available by the time you read this. Software ages as well, but at a somewhat slower rate. There will always be new features and updated ones added, inviting new customers to purchase and current customers to update. As long as the software does the job, some shops may work with the same version for many years.

A few years back, the subject of hardware consideration for CAM work could be suitable for several articles. Now, most concerns should focus on the processing power, RAM (temporary Random Access Memory) and the graphics card/display combination, even the hard disk capacity. In all cases, the bigger the specification number, the better choice. CAM software has been developed mostly for the Windows operating system, although some software does exist for the Apple operating system.

One piece of hardware — a CD or DVD reader/writer drive — is now part of every computer and serves as an inexpensive method of safely backing up valuable data. In addition, other inexpensive back-up accessories are available such as flash drives and external hard drives. If a direct connection between the CAM system and the CNC machine is desired, some CAM

A Reader for Programmers

systems offer a file transfer option. For a more complex dedicated DNC transfer, separate software is usually a better choice. Display is also very important — looking extensively at a computer screen can be tiring at best and may cause physical problems over time. Large viewing size and high resolution monitors are critical for comfortable work. Many graphics cards are made for gaming and may not be suitable for serious CAM work. Other features such as hard drive size are no longer a problem, as long as they are reliable and well maintained.

PROGRAMMING ENVIRONMENT

A typical working environment within CAM software can be easily summed up in three steps:
1. Toolpath Geometry
2. Toolpath Application
3. Program Generation

Depending on the software, many intermediate (or additional) steps may be available, but the three basic steps will always apply.

Toolpath Geometry

Toolpath geometry is a visual representation of the toolpath itself. It is always generated graphically, using points, lines, arcs, and other entities. True toolpath geometry is called an *associated geometry* — that means a change in either the toolpath or the operation will change all related data automatically.

Even the lowest priced (i.e., featured) CAM software offers the most common 2D toolpaths — drilling operations, contouring, and pocketing. More sophisticated software will also include special toolpaths and operations, such as helical interpolation, splines as part of the geometry, special cycles, subprograms, 3D machining, surfaces, and solids. When selecting CAM software, think also of future needs, such as the type of future work, as well as machine and control features.

CNC Tips and Techniques

Toolpath Application

Once the tools have been selected, a particular tool is applied to the selected geometry. The tool description not only identifies the tool dimensions, it also contains additional information, such as speeds and feeds, clearances, and offsets that are associated with the tool.

Program Generation

Generating a CNC part program is the last step in the process. In order for the process to match a particular machine and control system, the original processing has to be processed again — it has to be reformatted as per control specifications. This step is called post processing. A high quality post processor is one of the most important features of CAM programming software. The purpose of the post processor is to format the generic database into a part program that can be used on a CNC machine. CNC software vendors always include several post processors with their software. Keep in mind that these are generic post processors for the most common controls. In order to make them work in a specific way, post processors often have to be customized.

Some amount of customization is usually necessary. A qualified CAM programmer may optimize or customize a post processor or may ask the software vendor or an independent consultant to do it for a fee.

Customizing a post processor includes several steps, such as entering the known data about a machine tool, relevant data of the control system, such as format of the program, various commands and functions, use of the decimal point, etc. The program structure (or program template) is also part of a successful post processor. The format determines the order of tool sequences and machine activities (for example, the coolant).

A Reader for Programmers

Other Features

Combining part geometry with the tools used creates an operation. Operations are treated as units — within a unit, an individual change will update the whole operation, for example, changing a tool diameter, spindle speed, or a cutting feedrate. There could be a number of special features included in CAM software. One that is very important is the support for various cycles for drilling, milling, turning, grooving, and threading. Such cycles make manual programming a lot simpler and less tedious; having support for them in the CAM software makes it easier at the machine to make small machining changes. Some CAM software offers an open architecture model. In plain words, it means that anybody with the required skills can write special utilities that can be accessed by the CAM software. The ability to add external features greatly adds to the power of the CAM software.

Multi-Machine Support

Some machine tool manufacturers provide their own CAM software dedicated only to the CNC machines they make. Typical examples in this area are fabricating machines and some EDM machines. The software usually performs very well for the machine type supported, but it lacks flexibility to support other machines. It is quite suitable for a specialized shop, but not very practical for a shop that needs programming flexibility.

Most CAM software is integrated software, which allows the part programmer to use it for several types of machine tools, typically mills, machining centers, and lathes. The major advantage of integrated CAM software is its flexibility. Additional benefits may be more practical, but also important, for example, a similar interface for milling and turning shortens the programming time.

CNC Tips and Techniques

Text Editor

In many respects (and professional opinions), the text editor that is supplied with most CAM software should be only a viewer — it should not be able to make changes to the program. Of course, such changes are made every day because it is an expedient way to fix a problem, usually a minor one. The program generated by CAM software should not need editing; that is the whole purpose of purchasing it in the first place. Yet, many part programmers choose to make small changes manually. In principle, it is a wrong approach. The problem may have been fixed, but the program database, the part model, has not changed at all. It still contains the error. In an environment when several users work with the program, this approach can cause major difficulties.

At the same time, a text editor does have its own legitimate uses. For example, the editor can create and edit setup and tooling sheets, post processor templates, even configuration files. None of these applications will cause damage to the toolpath database. A text editor can also be used to print the program if hard copy is required.

CAD Connnection

A stand-alone CAM programming system does not require separate CAD software to define toolpath geometry. If a part drawing already exists (for example, if the customer delivers an Autocad drawing as a file), it is counterproductive to redraw the geometry. The CAM software should contain several built-in CAD conversion features. Autocad is still the leader in 2D drafting, and opens DWG drawings directly. DXF format is also available, but only suitable to translate basic entities. For 3D design, quality CAM software should convert the most popular CAD systems, such as Soiidworks, Inventor, Catia, and SolidEdge. Others are added frequently.

A Reader for Programmers

Managing a CAM System

CAM software requires a certain organization to support it; it needs planned strategies and focus. CAM system management establishes standards and procedures for all users, be it at the simple level of naming files to developing tool and material libraries, backup methods, and data security.

Training and Technical Support

Unfortunately, training is often neglected, even ignored. Professionally conducted training sessions will produce measurable positive results in a relatively short time. Most CAM vendors provide basic level training; some offer more specialized training. Initial training is usually general in nature, and should be provided to the person who has previously programmed manually. Any training should be practical in its orientation, preferably the usage of actual parts. A quality training program should also include short follow-up sessions that address problems encountered and answer many questions.

Technical support for CAM software is as important as for any CNC machine or control in the shop. A service or software maintenance contract is offered by many vendors, assuring of the latest updates and improvements, at a reasonable cost.

Readings From 2005

A Reader for Programmers

 ## Minimizing Program Length
May 2005, updated February 2013

In the past, the topic of handling long programs has focused on running such programs efficiently at the machine, during the actual part production with DNC support. Long programs are often the result of a CAM output, but they are also developed manually in many cases. Keep in mind that the word *long* is quite often relative to the amount of work involved and the available control memory capacity. CNC lathes, for example, have much smaller memory capacity than CNC machining centers. In either case, CNC programmers have several methods at their disposal that will make long programs shorter. Using the methods suggested in this column, it may often be possible to fit a long program into the control memory, without using the DNC method. Various methods are available to reach this goal and they can be adopted before the program is developed or after the program has been developed.

In CAM programming, the main key to quality program output is a quality post processor. The purpose of a post processor in CAM software is to provide a customizable platform for the desired program output. The same thoughts and efforts that go into post processor customization will also be considered in manual programming. There is no magic here — if you want to shorten the program, you must do it *without losing its integrity and purpose*. Removing an operation from a program will certainly make the program shorter, but the price may be too high to pay. When developing the CNC program, look at various features that offer shorter output right from the box, so to speak. They include various fixed cycles, multiple repetitive cycles, subprograms, macros, counters, and automatic corners. Starting with a program that is already as short as possible, it makes any subsequent effort that much more effective. Let's look at some options available.

CNC Tips and Techniques

Eliminating Characters

Once all possible methods have been used to make the program shorter from the beginning, there is only one more method left — *to eliminate all unnecessary characters* from the program. Yes, eliminating one character at a time will often work miracles on the final program length. You will need a good text editor, preferably one designed for editing of CNC files. Text editors offer feature called *mass substitution* (commonly known as the *find and replace feature*). Always work on a copy of the program, in case something goes wrong. Also, never make changes to the program that would negatively affect machining safety. Initial planning is important and the knowledge of program formatting is imperative. Here are the main areas that should be considered as methods suitable for the reaching the goal of a shorter program:

- Elimination or optimization of block numbers (sequence numbers)
- Removing program comments
- Removing unnecessary zeros
- Joining single-axis motions into multi-axis motions (if safety allows)

Block Numbers

Block numbers are almost always used for convenience, except in some special applications, such as multiple repetitive cycles or macro statements. Eliminating block numbers will make the most significant reduction in any program size. If you don't feel comfortable about eliminating all block numbers, use only one block number per tool, perhaps for the purpose of searching. Using block numbers in increments of one is more economical that common increments of five or ten, as fewer characters are stored.

A Reader for Programmers

Comments

If the control system accepts *comments and messages* within the program (those enclosed in parentheses), a great amount of available memory is used by them. Eliminating — or at least minimizing — the use of comments in a program will also go a long way to a shorter program.

Unnecessary Zeroes

Removing *unnecessary zeros* from the program may take a little bit more time and must always be done with care. Zeros that can be eliminated are leading and trailing zeros, as well as zeros programmed for convenience. For example, change all G00, G01, G02, G03, etc., to G0, G1, G2, G3, etc. Change full coordinate output to its minimal form, such as X0.1000 to X0.1 or even X.1 — all versions have the same meaning. Formats such as X1.0000 can be safely shortened to X1.0 or even X1. — all versions have the same meaning as well.

Combine Single-Axis Motions

Before *combining two or more single-axis motions* into one, always consider how such changes affect the machining safety. For example, it is quite common to program a tool approach along XY axes first than in the Z-axis, whereas tool return will be the opposite — the Z-axis first, followed by XY axes motion. Combining such motions will save only a few characters, but also may endanger machining safety.

Some additional methods may also be used. Shortening the length of CNC programs should always be considered a special situation, never the standard method of programming. There is nothing wrong with including convenience features in the program; such programs are easier to read, easier to interpret, and much easier to change. If the methods described are still not enough, remember, there is always the DNC method waiting in the wings.

CNC Tips and Techniques

 ## Conversion of Lathe Cycles
June 2005, updated February 2013

Most CNC lathe programmers would agree that the most useful features of a CNC lathe control system are the *multiple repetitive cycles*. Multiple repetitive cycles for CNC lathes have been an important part of control systems since the mid 1980s. Still, to this day, they present the most innovative approach of material removal, particularly in the areas of turning, boring, and threading. Over the thirty years of their existence, multiple repetitive cycles have gone through only two major changes. Earlier controls require these cycles to be programmed in a single block, later controls require two blocks of program input. This difference in programming method often presents a situation when one type has to be converted to another type — usually from a single block format to the double block format.

Converting Formats

To start, let's look at the word *convert*. Changing from one format to another is not a true conversion or — at least, it is not a complete conversion. The reason is that a double block format offers more features than a single block format. Also, keep in mind that you have no choice here; the control system determines the programming method. Typically, Fanuc control models 10/11/15 use a single block format, other control models (0/16/18/20/21...) use the double block format. What cycles are affected? All multiple repetitive cycles from G71 to G76 can be programmed in one or the other format, depending on the control. The finishing cycle G70 always uses a single block format.

Single Block Format

The single block format is the older of the two, and relies heavily on the settings of system parameters, generally inaccessible to the machine operator. I will use the most commonly used G71 and G76 cycles as examples in this column; other cycles fol-

A Reader for Programmers

low a similar pattern. The single block format of the roughing cycle G71 is:

G71 P.. Q.. U.. W.. D.. F..

In this single block format (spindle speed is assumed to be in effect), P and Q addresses refer to the block numbers defining the finish contour. U and W are specifications of stock amount left over for finishing, D is the depth of cut (written without a decimal point), and F is the roughing feedrate. In addition, some controls also accept I and K addresses that control the distance and direction of semi-finishing.

Double-Block Format

For controls requiring a two-block format, the G71 must be written at the beginning of each consecutive block:

G71 U.. R..
G71 P.. Q.. U.. W.. F..

The programmed data are similar but a bit more flexible. In the first block, the U address is the cutting depth (decimal point can be programmed), and the R address is the amount of retract from each cut. The second block has the same meaning as before — finish contour block number range P and Q, stock allowances U and W, and feedrate F. Apart from the more convenient way of programming the cutting depth, the addition of the R address represents the major change. In a single block format, the retraction amount was controlled by a system parameter; in the double block format, the programmer can specify such amount in the program directly.

G76 Threading Cycle

Even more profound change can be found in the threading cycle G76. In its single block format, the cycle uses the

CNC Tips and Techniques

following data:

G76 X.. Z.. I.. K.. D.. A.. F..

In this case, X specifies the final thread diameter, Z is the position of the thread end, I specifies the amount of taper (if used), K is the thread depth, D is the depth of the first pass, A is the thread angle, and F is the thread lead (feedrate). The two block version packs in a few more programmable features:

G76 P.. Q.. R..
G76 X.. Z.. R.. P.. Q.. F..

The P address in the first block includes the number of finishing passes, the length of the lead for pullout at the thread end, and the thread angle all in one. The Q address specifies the minimum cutting depth, and R is the finish allowance. In the second block, X and Z are the same as before, R specifies the amount of taper (if used), P is the thread depth, Q is the depth of the first pass, and F is the thread lead (feedrate). Neither P nor Q in the second block accept a decimal point.

As you see from the two examples, the main difference between the two cycle formats is the additional programmable parameters, which make the cycles much more flexible than using internal parameter settings. In either double block cycle, do not confuse the addresses in the first block with the same addresses in the second block. As you see, the *conversion* from a single block format to a double block adds certain features that are now programmable, offering more flexibility to the CNC programmer.

A Reader for Programmers

 ## Shifting Program Zero — Part 1
July 2005, updated February 2013

Program zero — also known as *part zero* — is the reference point of the part, also known as its origin. All programmed absolute dimensions are based on this key point. Typically, CNC programmers select the part origin partly for the convenience of programming, but mainly they have to consider the engineering intentions. Part dimensioning in the drawing indicates what sizes or locations are important and how they relate to each other. The program zero selection must always reflect that reality.

In the majority of programs, there is only a single program zero. In fact, there can only be one program zero at a time, but its location in the program can change; it can float. A typical example of such a situation is a standard bolt circle pattern located in a rectangular area. For the machine setup, selecting one corner of the rectangular block would be most convenient. On the other hand, selecting the bolt circle center would be more convenient for CNC programmers, as each hole location will be much easier to calculate. The question whether the corner or the center location should be selected as program zero can be answered quite easily — the corner can be selected for the operators, still allowing the programmers to calculate all hole locations from the center of the bolt circle. The secret is in two preparatory commands that the control system provides: G52 (local coordinate system) and G10 (data setting).

Surprisingly, although both commands are pretty much standard on most controls, they are often neglected by programmers. Let's have a look at each of them and see how they can make the job much easier for both the programmers and the operators.

The G52 Command

In this column, we look at the G52 command. By definition, G52 is called the *Local Coordinate System*. Whereas the pro-

CNC Tips and Techniques

grammed G54 work offset is used by the CNC operator to set the program zero, G52 is used to *shift* this zero to a different location, usually on a temporary basis. The real benefit is that the actual setting of G52 is controlled by the part program (the actual setting of the G54 offset is entered at the machine control, during setup). Using G52 does not need much effort, but three conditions are important for its successful use:

- G52 always works within the current work offset, such as G54
- G52 does not cause an axis motion — it only adjusts the current work offset
- The amount of zero shift used for the G52 must be known at the time of programming

The first condition explains the meaning of the G52 definition — *the local coordinate system is subservient to another coordinate system*. Programming G52 requires only the amount of difference between the program zero as determined by the work offset (G54) and the new location. For example, the **G52 X2.0 Y1.0** block in the program will shift the program zero two and one inches respectively in X and Y, into the positive direction from the *current* program zero (as set by G54). While the G52 is in effect, this new position is the program zero for all calculation purposes. To cancel the zero shift, just program a block containing zero values — G52 X0 Y0 — and the original program zero (as set by the G54 work offset) will return to normal.

Keep in mind that no axis motion will take place when the G52 shift is executed. All that happens at the control unit is the temporary updating of the internal work offset setting by the amount specified in the G52 block. Also keep in mind that because G52 is a programmable command, the amount of shift *must be known* at the time of programming. The primary usage of G52 is *within* a single part. It can also be used to shift zero between different parts located on the machine table, providing their exact locations are known. If the locations of other parts are random, use a different work offset, for example G54 for the first part, G55 for the second part, and so on.

The local coordinate system command G52 can be used on both machining centers and lathes. Its main benefit is that it allows CNC operators to set program zero in a more convenient or accessible location, and at the same time also offers CNC programmers a method of easier calculations of tool path coordinates. The result on both sides of the production process is minimizing possible errors, whether it is a programming error or an error that can happen during setup.

Shifting program zero by G52 command is not the only available method. Another command that can be used is G10 — *data setting command*. That will be subject of another essay.

 ## Shifting Program Zero — Part 2
August 2005, updated February 2013

A different essay described the G52 method of shifting program zero (part zero) in the CNC program. Now we look at another very powerful method, using the data setting command G10. The G52 command does not change the actual settings of offsets stored in the control system and can be used only when a work offset is in effect. In this respect, G52 is a true part zero shift command.

G10 — The Data Setting Command

Although G10 is also used to shift program zero, it can be used for changing *any offset,* not just the active work offset. In that sense, G10 offers many more possibilities and places more responsibility into programmer's hands. Regardless of its actual application, G10 will always change the current settings of the selected offset, with no means to undo, no means to go back to the previous settings automatically. This fact alone is a good reason to understand the data setting command well. All three offset groups can be changed by G10 — the work offset, the tool

CNC Tips and Techniques

length offset, and the cutter radius offset. If the absolute mode of programming (G90) is used, the offset affected by G10 will be *replaced*; if the incremental mode of programming (G91) is used, the offset affected by G10 will be *updated*.

The command G10 itself does not distinguish between the three offset groups — it needs additional parameters. For example, to select a work offset, the programmed entry will be **G10 L2 P.. X.. Y.. Z..** . Here, *L2* means an arbitrary work offset group number and the *P* address selects which one of the six standard work offsets will be affected. *P1* is G54, *P2* is G55, and so on. **X.. Y.. Z..** parameters contain the amounts that either replace (in G90 mode) or update (in G91 mode) the selected work offset. With tool length offset, the situation is similar. For example, **G10 L10 P.. R..** in the program will change the geometry offset setting of tool length offset specified by the P.. address. The R.. specifies the actual amount of change — again, replaced or updated. To make a similar change to the tool length wear offset, *L10* will be replaced by *L11*. To change the stored radius offset, *L12* or *L13* will be used, followed by the adjustment amount. There are several alternatives, depending on the exact control model.

As you can see, G10 can be a bit intimidating to new users, and is often considered an advanced command. To illustrate at least one example, let's go back to the zero shift in G54. The work offset is not normally known to the programmers, so the G10 will reflect the shift *from the current setting,* by a specified distance and direction (metric units):

Current G54 setting (unknown to the programmers):
X-400.000 Y-350.000 Z0.000
A programmed block **G91 G10 L2 X15.0 Y12.0** will change the current setting of G54 to a new setting:
G54 setting after G10 is processed:
X-385.000 Y-338.000 Z0.000

A Reader for Programmers

Although the programmers do not know the actual setting at the control, they know that the original program zero has been shifted by 15 mm and 12 mm along X and Y axes respectively. In order to return the G54 to its original settings, the G10 will include the reversed amounts:

G54 setting to the original program zero:
G91 G10 L2 X-15.0 Y-12.0

Shifting the program zero is not the only purpose of G10 — changing a radius offset through the program is also a very powerful application. Take, for example, a typical roughing and finishing contour tool path. Normally, two offset numbers will be used, and the CNC operator is responsible for both settings at the control. Using G10, the finishing tool path can include the stock amount, making it in effect the roughing tool path. For example, the current setting of radius offset D51 is 0.5000 (English units).

A program block **G91 G10 L13 P51 R0.03** will change the current radius amount setting of D51 to 0.5300, allowing a 0.03 stock for the finishing tool. Before the finishing tool becomes active, the block **G91 G10 L13 P51 R-0.03** will restore the original setting.

Whether the work offset, the tool length offset, or the cutter radius offsets are changed by the G10 command, the opportunities offered to imaginative CNC programmers are enormous. The purpose of these two essays was to encourage CNC programmers to look beyond the more traditional programming methods. Both G52 and G10 offer one great benefit — they increase control of the machining process by removing offset settings traditionally made by CNC operators, and placing them in the hands of CNC programmers.

CNC Tips and Techniques

 ## Threading Methods Compared
September 2005, updated February 2013

When it comes to single point threading on CNC lathes, part programmers have several choices — three, in fact. The oldest programming method was, and still is, a method that is also very time consuming. This is the original block-by-block threading, using preparatory command G32 (also known as G33 on some controls). Because of the many calculations required to prepare even a few simple threading passes, it did not take long for the control developers to make the programming process a little shorter, and the first threading cycle was born. This cycle is still known today by its preparatory command G92. Although a significant improvement for its time, the G92 cycle has not met the high expectations of part programmers and machine operators. Finally, Fanuc has come up with a true gem — the G76 multiple repetitive threading cycle. A bit of a mouthful for a description, G76 has quickly become the king of single point threading. How does that position the other threading methods, using the G32 and G92 commands? Is there still room for them?

The G32 Method

The G32 method of thread cutting is also known as the "long hand" method — and long it can be. The most fundamental understanding of G32 is that it is not a cycle of any kind. It is a plain programming of each and every step in the threading process. Take a typical single pass of a thread; it has four parts. From the start position, the tool will rapid to the diameter to be threaded, then the threading itself will take place. Next, the tool has to retract away from the thread and, finally, it must return back to the start position.

These four parts represent four blocks of the program. With seven passes, for example, the threading motions will only take twenty eight blocks of the program. For the best results, each

A Reader for Programmers

threading diameter has to be calculated individually, so the metal removal rate will remain fairly consistent. In addition, most threads benefit from a flank infeed, also known as angular of compound infeed. Thus, each start position will have to change its Z-coordinate, which has to be calculated trigonometrically, based on the depth of pass and the infeed angle desired. This could be quite a programming workout.

Although the G92 threading cycle made some progress towards a program that is shorter, it did not address any of the tedious calculations. G92 only removes the repetitive data from being written over and over again — that means you still have to calculate each threading pass diameter. In addition, G92 cycle is only suitable for straight or plunge infeed, not an angular infeed.

The G76 Threading Cycle

Introduction of the G76 threading cycle has changed the thread programming forever. In a very short program segment resides a lot of power — no more calculations of the number of passes or individual diameters, no more trigonometry to find the next Z-coordinate, plus a great control over cutting depth and infeed. All this and much more is squeezed into a one or two block segment of the program, for any number of passes. G76 cycle comes in two flavors: an older one that requires a single block of programming and a newer one that requires two blocks of programming. A lathe control system supports only one method or the other, but not both. Either method offers substantial flexibility at the machine, so the CNC operator can fine-tune the cutting conditions. Either method provides parametric specifications (meaning data can be changed externally rather than through the control settings). The two-block method offers more options relating to the finishing passes. Without a doubt, G76 is the cycle to use for most single point threading applications, perhaps even all threading applications.

Note the emphasis of the word most in the last sentence. G76 can do straight threads, tapered threads, multi-start threads;

CNC Tips and Techniques

it can do all external or internal threads. Yet, even G76 cannot be used — and is not meant to be used — for every thread type all the time. Extra deep threads, face or scroll threads, and certain special form threads cannot be programmed with G76 at all, or at least not to their full completion. G76 is mainly software driven. The human input is relatively minimal; the software takes most of the control. This is all great and works very well in the majority of threading jobs. What about the rest? If G76 cannot do a particular threading job in a satisfactory way, what is the alternative? Is there one?

There is only one answer to these questions: Yes, there is an excellent alternative and its name is G32. The simple cycle G92 just does not bring any significant benefits to be even considered. G32 on the other hand offers the absolute control over any thread the system software can handle. G32 is the ultimate solution, but there is a price to pay in terms of efficiency. Program using G32 must be right from the beginning. There is no fine-tuning possible at the machine; the program could be very long, and mistakes can squeeze themselves in very easily. In spite of all that, count on G76 as your everyday best choice for single point threading, but keep the door open for G32 — it may just save the day.

 ### When 1 Thou Equals 65 Dollars
October 2005, updated February 2013

One of my clients considers himself a master hole-maker and is proud of it, and quite rightly. He is at the top in all the latest drill technology, skills, and equipment. Making holes is his business in a very competitive field, and being number two in the area is not an option. Recently he was working on a part similar to tube sheet plates for heat exchangers — thousands of holes, quite deep. Watching the process was quite an experience and reminded me how other companies handle programs for such a large number of holes. Let's leave our master hole-maker

A Reader for Programmers

and have look at this situation in more detail.

G73 and G83 Cycles

Special drills do exist that allow drilling deep holes to the full depth without peck drilling on CNC machining centers. In the absence of such drills, the common practice is to program a fixed cycle — the standard G73 or G83 cycle — that is designed for such purpose. The G73 cycle is more suitable for breaking the chips because it does not retract above the part after each peck, only when the drill reaches the final depth. If you need to flush chips or get some more coolant into the hole, the G83 cycle may be a better choice.

On Fanuc and similar controls, either cycle requires the Z-axis start position (programmed as R), the final hole depth (Z), and the depth of each peck (Q). Although these cycles work well in most cases, they lack one feature. They are not able to decrease the pecking amount as the drill goes deeper into the hole. Optional Fanuc Macro B is an excellent method of programming for this purpose because it allows the user the exact control required by developing a flexible custom cycle.

What about those machine shops that do not have the macro option installed? Although there is no direct solution, there is a workaround of sorts. Just use the cycle two or three times at the same location, and decrease the Q amount for each run. A program may look something like this:

(G43 Z1.0 H01 M08)
G99 G73 X.. Y.. R0.1 Z-2.6 Q0.75 F10.0
R-2.5 Z-4.0 Q0.5
G98 R-3.9 Z-5.85 Q0.25
G99 X.. Y.. R0.1 Z-2.6 Q0.75 (ETC.)

There is a certain loss of time that can be significant when hundreds or even thousands of holes are machined with the same program, but it does the job without macros. Another pos-

CNC Tips and Techniques

sible problem encountered in machining a large number of deep holes has to do with assignment of the Q-amount. Generally, we do not think in small increments, and the Q is more often than not just a reasonable number, generally based on experience. Consider the following reasonable G83 cycle used for 5/8 drill with 0.87 peck depth, 4-7/8 part thickness, 0.06 breakthrough and 0.1875 drill point length (0.625 x 0.3). The programmed depth is 4.875 + 0.06 + 0.1875 = 5.1225:

G99 G83 X.. Y.. R0.1 Z-5.1225 Q0.87 F10.0

Drilling Thousands of Holes

The G83 cycle is necessary in this case to clean most of the chips out of the hole. This seems innocent enough; program data are definitely reasonable and, for a few holes, the program works well. Yet, it can be a costly piece of program code when, for example, 5000 holes are drilled. Let me do some math here.

To calculate the number of pecks, the control system divides the programmed pecking depth of Q0.87 into the total travel between R0.1 and Z-5.1225, a distance of 5.2225. This distance, when divided by 0.87, calculates 6.003 pecks, which in reality means seven pecks, as the control does not allow exceeding the Q-amount. That means six pecks will be done at 0.87 each, resulting in only 5.22 total distance traveled. The remaining 0.0025 will be the actual depth of the last peck. Because cycle G83 is in effect, the drill has to travel up by 5.22 and back by almost the same amount, it makes about 10.4 inches of travel in the air per hole, just to cut 0.0025. In addition, this 0.0025 is below the part anyway and more air will be cut.

Now, the costs. Take the 10.4 inches of extra non-cutting travel, multiply it by 5000 holes, and divide the result by the rapid travel of common 900 in/min rapid rate. The result is 57.8 minutes wasted, almost a full hour.

The solution is very simple. Just change the Q0.87 to Q0.871 (or even a little higher), which will eliminate the last peck. This

A Reader for Programmers

adjustment saves an hour of air cutting for every five thousand holes. Our master hole-maker would be pleased.

 ## Alternate CNC Machine Selection
November 2005, updated February 2013

There was a time when having a single CNC machine on the floor gave a machine shop a somewhat privileged status amongst its peers, even among customers and competitors — a pleasurable mixture of awe and envy. Being the first one to be in a possession of the greatest and newest technological marvel is still considered hip and cool today, even if CNC technology as such is not that new anymore. It is quite common to find even small machine shops with several CNC machines, typically purchased over the period of several years.

Of course, over several years, CNC technology has evolved in both the machine tool design and the control features. Having older and newer CNC machines working side by side presents several challenges. Apart from the standard scheduling challenges that apply to all machine tools in the shop, the CNC programmers and operators face challenges of their own.

Comparing CNC Machines

This essay looks at techniques that can be adapted in the part program and/or at the machine control. For example, let's use two similar CNC machining centers, *Machine A* and *Machine B*. They are both of similar capacity and features, so a certain job can be loaded on either machine. From a programming perspective, the major machine design considerations are the likely differences in the maximum spindle speeds and feedrates, capacity of the tool magazines, the tool change type, and spindle output ratings. Each directly influences the actual program development. Other possible differences, such as rapid travel rates, do not directly influence the programming method.

CNC Tips and Techniques

On the other hand, the control features may be a lot more significant, particularly if *Machine A* has a different control make than *Machine B*. For example, Fanuc, Yasnac, and Haas controls are programmed in a similar way, but they also have their own differences. The programmer has to know *all* features of both controls and consider *all* control options available.

Normally, a program is developed for a particular combination of machine and control. Being able to switch the job from *Machine A* to *Machine B* or vice versa for scheduling reasons, or if one machine is down, often means changing the program. Yet, for controls that belong to the same group (such as Fanuc, Yasnac, Haas, and many others), a program can be written in such a way that it reflects the common denominator of the machines and controls. It may not always be possible to switch machines without any program changes, but minimizing such changes goes a long way towards improvement in productivity.

Addressing Machine Differences

To address the machine design differences in the program, some ideas may be worth considering. When assigning tool numbers and their corresponding offset numbers for commonly used tools, assign those numbers that are common to both machines, typically in the low numbers. If the top spindle speed is required, consider programming the top speed of the machine that has the *lower* maximum. The same will apply for feedrates, although maximum feedrate is not commonly programmed and should not be an issue. Random type and fixed type of automatic tool change can be resolved in program structure (see below). Special features unique to one machine only cannot be compromised.

In programming structure, the first change may be right at the program beginning. When programming a startup block, its typical contents may be **N1 G20 G17 G40 G18 G49** or similar. Some controls require that the unit selection command G20 (or

A Reader for Programmers

G21) is in a separate block. Placing the G20 or G21 in a block by itself will work for all types of controls.

Similar structural change may reflect the tool change. Although it is customary to program **T02 M06**, for example, in the same block, some controls require that each command is in its own block. Again, using the two block method will always work. When using M-functions, most older controls only accept one in a block, newer controls may accept up to three. This convenience may work against the ability to select an alternate machine, so using a single M-function in a block provides a better program structure for that purpose.

There are other programming tools, some not always available. The most typical of them are the block skip function (slash) common to all controls, and the macro feature that is optional on most controls. Using a block skip method can be very powerful, particularly if the control allows the slash code in the middle of the block. In that case, for example, programming **S5000 M03 / S6000** in the same block will rotate the spindle at 5000 rpm if block skip is *ON*, and 6000 rpm if the switch is *OFF*. Note that both controls must have this special feature. Also note the inclusion of M03 *in front* of the slash.

Manipulating the program structure to achieve a certain result can be very useful. Ideally, programming for a single machine/control combination is preferred. Selecting an alternate machine because of an emergency will generally justify the changes required in the program. Selecting an alternate machine for scheduling reasons requires a certain compromise in programming and even machine setup. Understanding the possibilities will result in better decisions.

CNC Tips and Techniques

 ## Standard and Rigid Tapping — Part 1
December 2005, updated February 2013

In this two-part essay, we look at the two common methods of producing tapped holes on CNC machining centers. In this first part, the focus will be on tapping using floating tap holders whereas the second part will cover the subject of rigid tapping.

Tapping operation on conventional mills, lathes, and CNC machines has been one of the basic operations to machine holes. To this day, with all its disadvantages, one of the most common methods of tapping uses special tapping heads, known as the *tension-compression* type. Another term for this type of tool is a *floating tap holder*.

Standard Tapping Method

In a typical tension-compression tap holder, the tap is held firmly so it cannot fall out, but loose enough to provide axial flexibility upon entering a drilled hole. A thread in the hole is formed by the size and shape of the tap. Although still very popular, floating tap holders present one major problem — they make it difficult to maintain precise thread depth, particularly in blind holes.

Basic Principles

The main principle of standard tapping is the synchronization of spindle speed and tapping feedrate. That is achieved by multiplying the spindle speed (r/min) by the tap lead (pitch), resulting in the feedrate per minute. At the beginning of the cut, the tap has to accelerate from zero feedrate to a rather heavy feedrate. The time required to reach the programmed feedrate is called *acceleration*. In order to reach the programmed feedrate before any contact with the material, the tap must start further away from the hole — this is the R-level in fixed cycles. As the tap enters the hole, it encounters a resistance, and the tap is pushed slightly into the holder (using its compression capabili-

A Reader for Programmers

ties). When the tap reaches the hole bottom and the spindle stops, the deceleration of the spindle forces the tap to be slightly pulled out of the tool holder (using tension of the holder).

Standard tapping is almost always programmed using one of two tapping cycles: G84 (right hand tapping) and G74 (left hand tapping). As you will find in the second part of this essay, many control systems support the same format for rigid tapping either as provided or slightly modified. The major difference is not much in the programming technique as it is in machine design and suitable tooling.

Why Underfeed?

Most experienced CNC programmers have faced various tapping problems in their careers. The floating tap holder is an imperfect tool to say the least. It is a mechanical device and, as such, subject to some variables, including spindle fluctuations. If the tapping operation is not successful the first time (assuming the program is correct in principle), programmers often use a simple technique for a quick fix: *tapping feedrate reduction*. This reduction is called *underfeeding* because it is lower than the normal tapping federate, which is always r/min x pitch. Here are the reasons why this technique often works.

Most spindles may fluctuate up to ±10 r/min, particularly at high speeds. Because the tapping feed is always a direct relationship between the spindle speed and the tap pitch, any spindle fluctuation throws this relationship out of balance. When the tap in a floating holder enters the hole, it is pushed a bit into the holder, before the tap has a chance to "bite" into the material. This is the *compression* part of tapping. When a tap reaches the Z-depth, the spindle stops, but not right away; it has to wind down first. In order not to break the tap, the floating holder allows the tap to be pulled slightly out of the holder by its own force, after the Z-axis motion stopped. This part is called *tension*. Compression means pushing; tension means pulling. Now back to the spindle fluctuation.

CNC Tips and Techniques

Take a floating tap holder and try to push, then pull, the part that holds the tap. It is much easier to pull it out than push it in. In technical terms, it means tension presents fewer, if any, problems. In order to eliminate thread damage upon tap entry or exit, the feedrate can be reduced by 2% to 5%, anywhere between 98% and 95% of the nominal tapping feedrate. In this case, the tap is pulling from the tap holder and compensates for the spindle speed differences.

Keep in mind that a slight underfeeding when using floating tap holders is not always a solution to better tapping, but it may be the quickest solution in certain cases.

Floating tap holders are on the way out. Although they have served the industry for many years, a newer technology has become the standard. Since the introduction of rigid tapping in the early 1990s, the tapping methods and techniques have changed dramatically. The column covers the subject of *rigid tapping*.

Readings From 2006

A Reader for Programmers

 ## Standard and Rigid Tapping — Part 2
(* Part 1 see Dec. 2005)

January 2006, updated February 2013

Part 1 of this two-part essay detailed tapping methods that use a floating tap holder to produce a thread in drilled holes. Now we look at the other method: *rigid tapping*.

Rigid tapping is a common description of a technology known as *synchronized tapping*. Most CNC manufacturers offer the rigid tapping method as a standard feature on their machines. It is important to realize that the machine tool has to be designed for such purpose, including the control system. Rigid tapping must be supported by the machine design and related software.

Basic Principles of Rigid Tapping

The most visible difference between standard and rigid tapping is the tool holder. Whereas standard tapping requires a spring loaded floating tap holder, rigid tapping can use any solid tool holder designed for end mills. The reason why a floating tap holder is not required in rigid tapping is that the spindle itself does all the required synchronization via a built-in encoder. Tapping using the synchronized method also uses fixed cycles of the control system, often the same ones as for standard tapping. It is the computer software in the control system, not the cutting tool, that synchronizes the programmed spindle speed (r/min) with the programmed tapping feedrate (mm/min or in/min) to match the thread lead (pitch).

Benefits

If there is only one benefit of rigid tapping over standard tapping, it is the depth control. Unlike tension-compression holders, the machining power for rigid tapping is provided by the spindle, rather than torque of the cutter. Because any tool wear requires higher torque, the tapping depth suffers as a

CNC Tips and Techniques

result. This is not an issue with rigid tapping. In fact, the horsepower (or kilowatt) rating of the machine tool is the only limiting factor.

There are other benefits as well. The rigid tapping method is more suitable for making threads in tough and hardened materials, as well as in deep holes. Some space age and other exotic materials may even be tapped using the pecking method, common to drilling deep holes. In this case, a special part program will have to be developed, but that is not a difficult task. Re-tapping is also possible, as long as the starting point and depth never change. This is a benefit for holes that require a certain tolerance on the tapped depth. Overall, a significant reduction of cost per hole can be achieved with rigid tapping over tapping with floating holders.

Setup

As solid holders are used, it is very important to make sure the tool deviation from the spindle centerline is at absolute minimum, and does not exceed 0.005 mm (5 microns or about 0.0002 inches) of total indicator reading (TIR). Equally important is using high quality taps, which are less likely to run-out. They may cost more than standard taps, but are more economical overall. Large run-out will cause an oversized thread and may even break the tap, particularly for small hole sizes.

Possible Problems

Although rigid tapping is supported by the majority of modern machining centers, there are some differences in design. It would be a mistake to assume that synchronization of every machine spindle is perfect and that a solid tap holder solves all tapping problems. Various physical factors contribute greatly to the quality of the thread tool life, particularly in volume production. No significant difficulties are generally encountered in small batches, but tap wear over many holes has to be considered as a serious potential problem because the quality of tapped holes will gradually deteriorate. The reasons for reduced tool life are mechanical forces applied mainly during the Z-axis motion. There is also the issue of efficiency when using rigid tapping for a large number of holes.

Programming Approach

As in every effort to develop programs for specific machine tool features, rigid tapping also requires good knowledge of machine specifications and control features, including programming structure and format, mainly related to spindle speed and tapping feedrate.

Spindle speed is always a matter of selecting the most suitable speed within a certain range. For any machining operation, the material being machined is a significant factor influencing the spindle speed. In rigid tapping, there is another factor to consider — because the spindle speed is directly related to the tapping feed, a constant torque range of the spindle motor is required. General recommendations vary somewhat between machine manufacturers, but the general consensus is that rigid tapping should not exceed 2000 r/min (higher is also possible), depending on material. Feedrate selection is the same as for the older tapping method. Rigid tapping feedrate must always be the exact result of *r/min x pitch*, with no underfeeding.

Special Functions

If a standard G84 or G74 cycle can be used for both floating and rigid devices, the question is how does the control recognize which one is the programmer's intent? The answer depends on the manufacturer. In fact, some manufacturers do not allow G84/G74 for rigid tapping at all, and provide a special G-code instead. Others allow G84/G74, but also require a special command or function to be used before the tapping cycle is used. Always consult the machine manuals for exact program format. Rigid tapping is mostly a machine function and the additional programming requirements are minimal.

CNC Tips and Techniques

 ## Mastering M-Functions
February 2006, updated February 2013

Every CNC program has them. They appear in the program at various places and they have a lot of potential and they certainly have a lot of power in the program. They are called *M-functions* — special program words beginning with the letter M. For many programmers, they are at the same level of importance as the venerable G-codes (preparatory commands). Yet, there is a difference in their applications. Whereas many preparatory commands are pretty much the same for the majority of control systems, the M-functions do not share this distinction as readily.

Defining M-functions

M-functions are typically defined as *machine* functions or, more accurately, as *miscellaneous functions*. Their purpose in the program is two-fold. In one group, they control certain flow of the program, for example, functions such as M00 (program stop), M01 (optional program stop), and M30 (program end), as well as the subprogram related functions M98 and M99. The other group is much richer. It contains the true machine functions; they relate to the operation of various machine tool features and accessories. Their list can be quite long, depending on the machine type and design. Typical common functions of this group are those controlling coolant activity (M08, M09), spindle rotation (M03, M04, M05), tool change (M06), and many other standard machine-related functions. Virtually all CNC machining centers require these features and that makes these functions pretty consistent between controls.

Also related to machine design are M-functions that are *unique* to a particular manufacturer, even if they serve the same purpose. For example, table clamp/unclamp functions for one horizontal machining center may be different from table clamp/unclamp functions of another horizontal machining center. Depending on the number of special features, some complex

A Reader for Programmers

machining centers may have dozens or even hundreds of M-functions available; some even use three digits rather than the common two digits. Most of them are the ON/OFF type — active or inactive state. For these machines, the CNC programmer has only one resource, and that is the manufacturer's documentation.

Programming Perspective

From the programming perspective, miscellaneous functions are pretty flexible to work with. Two areas of focus are those M-functions that should be programmed in a separate block, with no other data, and the behavior of the function when programmed together with an axis motion. Some controls require program end functions (M02, M30, M99) to be in a block of their own. Even if such a rule does not apply, many programmers follow this format anyway because it makes the program easier to interpret. The same approach applies to M00 and M01 program stop functions. When it comes to functions that are machine oriented rather than program oriented, some significant differences should be expected. A typical example is the M06 automatic tool change function. This function is often programmed together with the tool number, for example, T05 M06. Some controls accept the two commands together in the same block (regardless of their order). Others require the tool command in one block, followed by the tool change command in the next block.

Number of M-functions

Another important feature is the number of M-functions in one block. Not many control systems allow more than a single M-function in one block. Others offer up to three such functions in a block, providing they are not in conflict with each other. The best strategy is to program only one M-function per block, to prevent possible syntax error at the machine. Such a strategy also makes the program more portable.

CNC Tips and Techniques

Behavior of M-functions

Finally, a very important subject relating to M-functions considers how they behave when programmed with some machine activity, typically with an axis motion. Whatever the exact activity, common sense prevails. Take for example, G00 X20.0 S1200 M03. The spindle speed starts *simultaneously with the motion*, because it makes sense. A similar block G00 X20.0 M05 will force the spindle to stop *after the motion has been completed* — again, it makes sense.

There are times when such an approach is the only way to guarantee a safe cut and preserve the program integrity. For example, in tapping on a CNC lathe, the G32 command is typically used, followed by the Z-axis position of the thread end and the feedrate. The only correct programming is to include the M05 in the same block as the motion, not in the block that follows, for example, G32 Z-0.75 F0.05 M05. Why? In a single block operation, failure to do that would cause a stripped thread or even a broken tap while the spindle is still rotating once the block processing has been completed. Because the tap has to reverse before moving back to the start position, it requires M04. Again, the only correct programming approach is to include the M04 together with the return motion, for example, (G32) Z0.4 M04. During the motion into the part, the spindle will stop *after* the motion is completed. During the reverse motion, the spindle starts rotating *simultaneously* with the motion.

Miscellaneous functions are quite easy to program, but there could be situations when a little more in-depth knowledge may save the day — or at least the part.

A Reader for Programmers

 ## Tool Length Setup — Three Methods
March 2006, updated February 2013

For a typical setup of a vertical CNC machining center, it is necessary to provide the control system with the exact location of the part zero, measured from the fixed machine zero position (home). Although this procedure is fairly straightforward for the X and Y axes, setting the Z-axis measurements is a bit more involved (unless your machine supports automatic tool length setup). For the XY axes, a simple edge finder or a corner finder is used during setup and typically remains the same for all tools. The measured dimensions are stored as work offset (G54–G59). In order to set the Z-axis, the length of each tool has to be known and registered as tool length offset. As an example, for a single tool, the Z-axis in the work setting can also be used and called by the G54 command. As most CNC programs use more than one tool, this method is not possible, and the tool length offset (G43) must be used instead. Registry settings allow entry of each tool length separately, and they can be called by their number, using the address H in the program. For example, G43 H03 will retrieve tool length offset number three.

At the machine, the CNC operator may use one of three common methods to set the tool length offset:

- Preset Method
- Touch-Off Method
- Longest Tool Method

Preset Method

This is the oldest and still commonly used method of presetting the tool away from the machine. It requires an external tool pre-setter and a person who sets the tool length for each tool in the program. Once the tools are set, they are assigned to the CNC machine and loaded into magazine. Because their dimensions are known, all the operator has to do is to enter them into

CNC Tips and Techniques

the offset registers; even that can be done automatically, with data stored on special chips, built into the tool holder. The preset dimensions are actual tool lengths, measured from the tool tip to the spindle gage line, and are typically stored as positive values. Overall Z-axis distance of the machine tool must also be known.

The obvious advantage of this method is very little time loss during the actual setup. A less visible benefit is that presetting tools off-machine provides a better environment for tool management. The main disadvantage is its cost, not only in the initial equipment, but also in its operation. Although any shop can use the tool presetting as a method of choice, it remains the domain of larger companies or companies that run large batches of the same part.

Touch-Off Method

Many small machine shops, and even more job shops, use a method of touching the tip of each tool at the surface that represents the Z0 in the program. The measured dimensions are not the actual tool lengths, but individual distances from the tool tip to the part Z0, measured from the Z-axis home position. These settings are typically entered as negative values and require no other measurements or settings.

Although the main advantage of this setup method is in its simplicity, the main disadvantage is that it often creates a long gap between two jobs, resulting in unproductive time. It might be justified in a non-production oriented environment, such as tool and die shops, prototype work, etc.

Longest Tool Method

The last available method is one that also uses the touch-off procedure, but with a twist. Rather than setting the length offset for each tool relative to the Z0, only one tool will be set to Z0. All other tools will be set relative to the tool point of this master tool. From a practical point of view, the master tool should be the longest tool, mainly for reasons of adding safety into the setup.

A Reader for Programmers

Also, if the master tool is the longest tool, setting values for all other tools will be positive. The longest tool that is most suitable is not an actual cutting tool, but a solid rod with a rounded end, permanently mounted in a tool holder. The cost of the extra holder is minimal comparing to the savings in turnover between jobs. Of course, this tool does not have to be mounted in the magazine all the time — only for setup. When a new job comes in, the operator sets the tool length of the longest tool, and all other tools fall in line. The longest tool is stored as the Z-amount from the work offset screen; other tools are set the normal way, in the tool length offsets registers.

One disadvantage of this method is that it benefits only jobs where the majority of tools remain the same. For jobs that require different tools, this setup method may not be the most practical.

Regardless of the method used, the programming format does not change. The required settings are made entirely at the CNC machine. A typical program entry may have a block N3 G43 Z0.1 H01 M08 or one similar. Some programmers like to use G49 for canceling the tool length offset. This is not necessary; any machine zero motion along the Z-axis will cancel the length offset automatically.

Selecting the most suitable setup method for your shop will make a big difference in a short time.

(Extreme) Power of Subprograms
April 2006, updated February 2013

Most CNC programmers have used subprograms ranging from a simple application to some very complex machining. For the newcomers to the field of CNC programming, a subprogram is a program that contains repetitive data, for example, a common pattern of holes or common machining motions. Subprograms are stored in the control system separately from the main program. Rather than writing such a pattern for each

CNC Tips and Techniques

tool, the main program calls the subprogram and retrieves the hole locations for any tool. This method results in significantly shorter programming times, makes the program easier to understand, and simplifies editing, if necessary. The M98 Pxxxx command is used to call a subprogram previously stored in the control as Oxxxx. The M99 function is used to end the subprogram, to distinguish it from the M30 function, which ends the main program. Once the subprogram is completed, the program processing normally returns to the block that immediately follows the subprogram call.

Using Subprograms — An Example

Just how far can this method of programming be pushed? Let's have a look at a really extreme situation, never mind how improbable — and even impossible — it is. In just twelve program blocks, a square grid of *one hundred million* holes will be programmed, using only a single tool — a spot drill — as an illustration of this powerful method.

The grid should be easy to visualize. It is made of 10,000 rows of 10,000 holes each, a total of 100,000,000 holes, arranged in a square pattern. The spacing between holes will be 4 mm along the X-axis and 4 mm along the Y-axis. Depth of cut will be 1.5 mm, starting 1 mm above the part (the R-level), for the total cutting motion of 2.5 mm. The lower left hole is located at X10.0 Y10.0 and will be machined first. Both the main program and the subprogram program will use a feature of the control system allowing *repetition of an increment* between holes so many times, as well as a *repetition of the subprogram* execution, using the address *L* (or *K* on some controls). Here is the main program and subprogram listing:

```
(MAIN PROGRAM)
N1 G21 G17 G40 G80
N2 G90 G54 G00 X10.0 Y6.0 S1500 M03
N3 G43 Z1.0 H01 M08
```

A Reader for Programmers

```
N4 G99 G82 R1.0 Z-1.5 P100 F150.0 L0
N5 M98 P7001 L5000
N6 G80 G90 G28 Z1.0 M09
N7 M30
%

O7001 (SUBPROGRAM)
N101 G91 Y4.0
N102 X4.0 L9999
N103 Y4.0
N104 X-4.0 L9999
N105 M99
%
```

That's all there is to it. In just twelve blocks, one hundred million holes have been spot drilled. Now, some necessary explanations are needed. Although the first hole is at X10.0 Y 10.0, in the block N2, the Y-position is 4 mm lower than the first hole. There is a good reason for it. The subprogram makes two rows of holes, so it twice has to contain the 4 mm Y-axis shift between rows. As there is no shift for the first hole, a dummy position had been established. When the subprogram is called, the L0 address guarantees that this dummy hole will not be machined, but all fixed cycle data will be preserved for all subsequent holes. Block N5 calls the subprogram O7001 and repeats it five thousand times, two rows at a time. Within the subprogram, the L9999 specifies that there are 9,999 spaces to move. The repetition in incremental mode always means *how many times*, not *how many holes*.

Now, for the extreme part of the main title. If you had to make a time study for such a job, what will be the cycle time for this program (just for the spot drill)? Let's look at its individual elements only. First, the rapid motions will be based on 25,000 mm/min rapid rate for all axes, so the 4 mm spacing for 100,000,000 holes will require 16,000 minutes or 266.67 hours or

CNC Tips and Techniques

11.11 days to complete. As each rapid retract from depth is 2.5 mm, it will take 10,000 minutes or 166.67 hours or 6.94 days to perform these motions. The spot drill will pause (dwell) for 0.1 second at the bottom of each hole, for the total of 166,666.67 minutes or 2,777.78 hours or 115.74 days. Finally, for the actual cutting motion, the total travel is 2.5 mm per hole at 150 mm/min. That motion totals 1,666,666.67 minutes (or 27,777.78 hours or 1,157.41 days) to finish.

What is the grand total? 1291.20 days or 3.54 years of spot drilling only, with absolutely no interruptions of any kind. Of course, there is no CNC machine on the market that could accommodate such a large piece of material, and the program as shown is indeed a very extreme and practically impossible example. Yet, it shows the power of repetition combined with a clever design of both the part program and the subprogram.

 ### Special Purpose G-Codes
May 2006, updated February 2013

Anybody even loosely associated with CNC work is probably familiar with the G-codes, at least with their existence. *Preparatory commands,* as they are officially called, have a single purpose — they give a specific meaning to other commands in the part program. For example, a programmed command X10.0 can be interpreted as 10 mm, 10 inches, or even as 10 seconds. If this command represents a dimensional program entry, it could be an absolute position or an incremental distance and direction, in rapid or feedrate mode. The exact meaning is defined by one or more G-codes. For example, an incremental feedrate motion of 10 mm will require that G21 (metric mode), G91 (incremental motion), and G01 (linear interpolation) commands are in effect, along with appropriate feedrate.

Many preparatory commands are well known to CNC programmers and operators alike, as they are used in virtually every

A Reader for Programmers

program, such as those shown in the example. Yet, a list of G-codes for a particular control can be quite long and contain many that are hardly used. Typically, a preparatory command will have two digits following the letter G. Three digit G-codes generally identify the G-code as a particular cycle or function to the CNC machine that uses them; they are not usable anywhere else. Three-digit G-codes are often special macros supplied by a particular machine tool builder. Within the two-digit category of G-codes, there are also several that are fairly common, yet hardly ever used in a part program.

G22/G23

One pair is **G22/G23**, defined as *Stored Stroke Check*, ON and OFF respectively. In plain terminology, using these commands allows the programmer to define special three-dimensional zones either as ones that allow a tool from entering or as zones that prohibit a tool from entering. Such definitions may prevent a collision with a certain part of the machine or a fixture.

G25/G26

A relatively recent pair of other G-codes, **G25/G26**, is used to detect severe spindle fluctuation caused by heat. When *G26 Spindle Fluctuation ON* is in effect, the control system monitors whether the fluctuation is within specified tolerance range or not. Should the tolerance be exceeded, the control will issue an alarm, warning the user. In order to use this function correctly, a consultation with a service technician is recommended.

G27/G28/G29/G30

Another set of G-codes is a foursome related to the return to machine zero: G27/G28/G29/G30. Whereas the G28 command and, to a large extent, the G30 command are used to move the specified axis or axes to the machine zero (G28 to the primary and G30 to the secondary machine zero), the G27 and G29 are

CNC Tips and Techniques

seldom, if ever, used in a typical part program. Command G27 is a machine zero position check. It is programmed with one or more axes and checks whether the position specified corresponds to machine zero position. If not, the program processing stops and alarm is raised by the control. Its practical application has ceased with the demise of G92 (G50) programming *(position register)*. There is no need for this command in a program that uses work offsets (G54 to G59 or higher). G29 is the opposite of G28/G30. Whereas G28/G30 represents a motion to machine zero via an intermediate point, G29 is a *motion from machine zero to a specified position via an intermediate point.* This command is virtually useless in typical programs.

Additional Codes

Another virtually useless command is G44 *(tool length offset negative)*. It may have its adherents, but there is nothing that the ubiquitous G43 command *(tool length offset positive)* cannot achieve.

A command that may be somewhat of a puzzle is G31, defined simply as the *Skip command.* This command is indeed a very special one because it is exclusively used with probing on CNC machines (in-process gauging). Its purpose is to move the probe from a clear position, at a feedrate (just like G01), to a position to be probed. Once the probe detects a contact with the material, the motion stops and the remainder of the motion is not completed; it is *skipped,* hence the command name.

G6x Codes

Finally, there are several G-codes in the G6x range, namely **G60** *(Single direction positioning),* **G61** *(Exact stop check mode),* **G62** *(Automatic corner override mode),* **G63** *(Tapping mode),* and **G64** *(Normal cutting mode).* The default is **G64**, normal cutting mode.

G61 is the modal version of **G09,** and can be useful for increased precision when a tool has to move around a sharp corner at a high feedrate, particularly if the corner forms a narrow

A Reader for Programmers

angle. G62 may improve the surface finish when used to cut sharp inner corners in cutter radius offset mode. Its function is to automatically adjust the cutting feedrate. A tapping mode programmed with G63 disables the feedrate override and the feedhold button, and G64 returns the cutting motions to their original programmed mode. Each mode selected replaces any other mode.

G60 stands separately. Its purpose is to force a unidirectional tool positioning, regardless of which direction the tool comes from. It can be used to compensate for some minor backlash, but it does *not* remove it. Backlash problems require qualified service technicians.

There may be other G-codes not commonly used, but the ones mentioned in this column should provide some understanding of the more common special purpose G-codes.

Well-Structured Program Structure
June 2006, updated February 2013

A fairly experienced CNC user asked me recently, *"Are there any rules as to how actually write the program?"* His interest was in the format of the program, not its actual contents. He was interested if there are any preferable ways to structure a part program to gain certain advantages. Indeed, what is a good structure? In fact, what is a program structure in the first place?

A typical CNC program will contain several tools, arranged in the order of machining. Each tool will be programmed in three unique sections: 1) a few blocks at the *beginning* to move the tool to a cutting position, 2) the actual *machining* section, and 3) another few blocks to move the tool away from the work to make room for the next tool. No doubt, the machining section — the actual toolpath associated with cutting — is the most important one and is generally immune to any particular format. The beginning and the end of each tool are the areas where many

CNC Tips and Techniques

structural improvements can be made, mainly in the beginning section.

A Sample Milling Program

Consider a typical start of a CNC program for milling:

```
N1 G21 G17 G40 G80 G90 G54
N2 T01 M06
N3 G00 X.. Y.. S.. M03 T02
N4 G43 Z.. H01 M08
  <... machining with tool T01 ...>
```

The program looks all right, containing all data required at *the start. Yet, a few changes in the data order* — in the structure — will make a significant difference in case a problem does occur. First, the G21 (or G20) command should be in a block of its own. Even if the control allows the units selection combined with other G-codes, programming this command in a separate block will make the program more portable between machines with similar controls.

The same applies to the tool change. There is nothing wrong with T01 M06 in one block, but splitting this format into two blocks makes the program more flexible. The first block is often called the *safety block*. The idea behind the expression is to store all required settings and/or cancellations first, so the program starts safely, without any defaults or leftover settings from previous programs.

Further Improvements

Including such a block at the top of the program is recommended; however, the structure as presented is far from the best. Consider this unwanted, but common, possibility. During machining, the program switches to G91 incremental mode and, shortly after, the control system generates an alarm for some reason. When the alarm is cleared, the operator returns to the block

A Reader for Programmers

after the tool change and repeats the toolpath, this time in incremental mode. The reason? Command G90 was programmed too early, *before the search block*. Exactly the same problem may happen if G54 was changed to G55, and the program for a particular tool had to be repeated. These are not programming errors per se, but actual situations that can cause an error during machining. All it takes to make the program better is to *change its structure*. It will also be easier for the CNC operator to work with such a program:

```
N1 G21
N2 T01
N3 M06
N4 G17 G40 G80
N5 G90 G54 G00 X.. Y.. S.. M03 T02
N6 G43 Z.. H01 M08
  <... machining with tool T01 ...>
N15 G00 Z2.0 M09
N16 G28 Z2.0 M05
N17 M01
```

Selecting the units mode in the first block is normal, as units should never be changed within the same program.

Another common problem CNC operators often encounter is when a certain tool has to be repeated *after* another tool is already in the spindle. In such cases, all it takes to eliminate a manual tool change at the machine is to repeat the current tool number *at the beginning of each tool:*

```
N18 T02  (TOOL CALLED IN N5 IS REPEATED)
N19 M06
N20 G90 G54 G00 X.. Y.. S.. M03 T03
N21 G43 Z.. H02 M08
  <... machining with tool T02 ...>
N35 G00 Z2.0 M09
```

CNC Tips and Techniques

N36 G28 Z2.0 M05
N37 M01

Note the repetition of the G90 G54 G00 commands in block N20, *after the tool change.* Strictly speaking, none of these commands are required unless one or more had been changed earlier. Their repetition is intentional in order to achieve a better program structure, which translates into more efficient machining. Programming the optional stop at the end of each tool further improves the program structure.

These suggestions are not only aimed at the CNC programmer who develops programs by manual means, but also for the CAD/CAM method of programming. Virtually all CAM software provides a customizable post-processor that formats the program output. Incorporating a proven program structure into the post-processor will result in consistent part programs that are easy to interpret and navigate.

 Imagining a Mirror Image
July 2006, updated February 2013

As a standard control feature, a mirror image is quite simple to use but also requires a certain amount of special care. Although most CNC programmers are familiar with mirror images in general, the type of jobs the programmers do may not be suitable for its application. What is there to know about mirror images?

In some books or training materials, a mirror image is often described as *axis inversion or axis symmetry.* Although not incorrect, this term describes only a portion of what a mirror image does. As the name of the function indicates, a part programmed in one quadrant can be flipped — *mirrored* — into another quadrant. Depending on the quadrant in which the mirroring takes place, the function changes not only the axis direction (positive to negative and vice versa), but also the direction of an arc

A Reader for Programmers

motion (G02 to G03 and vice versa).

Considered separately, neither change causes any machining problems. That is not true about the third change. Changing direction of an axis also changes the cutter radius offset from G41 to G42, which means from climb milling to conventional milling, when right-hand tools are used. This last fact is the one to consider seriously before using mirror images.

For example, in a standard layout of X and Y axes, Quadrant I is the upper right part of the grid, Quadrant II is at the upper left, lower left is Quadrant III and lower right is Quadrant IV. If the original toolpath is developed in Quadrant I with G41 (climb milling) in effect, the climb milling mode will be maintained in Quadrant III, but will be changed to G42 (conventional milling) in Quadrants II and IV. Depending on the job requirements, this may present tool life and part quality problems.

Types of Mirror Images

The main benefit of mirror image is in time saved during program development. From a technical viewpoint, there are two types of mirror image available:

By *setting* — the mirror image is set at the machine control
By *program* — the mirror image is programmed with miscellaneous functions

In both cases, the program has to be written in such a way that allows the use of mirror images. Keep in mind that once the function is activated, all XY motions will be mirrored, including all rapid motions (the Z-axis is not mirrored). The only exception is machine zero return, which is not affected by any mirror setting. It is important to start and end a toolpath at the same point, in order to make mirroring successful. Program features not related to axis motion are not affected, for example, spindle rotation.

CNC Tips and Techniques

By Setting

Setting at the machine is usually done through the SETTING screen, and has options for each axis (*X* and *Y*) to be 0:OFF or 1:ON. The CNC operator sets the required mode before repeating the part program in different quadrants.

By Program

A programmable mirror image is far more common, as it provides better control of the program flow. Typically, the machine tool manufacturer provides three (sometimes only two) miscellaneous functions (M-functions) to cover all setting possibilities (check your manual). For example, a particular machine uses M21 to activate the *X*-axis mirror, M22 to activate the *Y*-axis mirror, and M23 to cancel the mirror image in either axis. Programmable mirror images are commonly used with a subprogram, and turned ON or OFF before the subprogram is called. Here is a small example of a 45-degree motion between two points. The slot is programmed in Quadrant I and mirrored to Quadrant II, Quadrant III and Quadrant IV. The common start/end point to all four quadrants in this case is X0Y0:

```
O8001 (SLOT SUBPROGRAM)
N101 G00 X70.0 Y50.0
N102 G01 Z-5.0 F125.0
N103 X170.0 Y150.0
N104 G00 Z2.0
N105 X0 Y0
N106 M99
%


O1001 (MAIN PROGRAM)
N1 G21
N2 G17 G40 G80
N3 M23              (MIRROR MODE OFF)
N4 G90 G54 G00 X0 Y0 S1200 M03
N5 G43 Z2.0 H01 M08
```

A Reader for Programmers

```
N6 M98 P8001      (QUADRANT I)
N7 M21            (X-AXIS MIRROR ON)
N8 M98 P8001      (QUADRANT II)
N9 M22            (Y-AXIS MIRROR ON)
N10 M98 P8001     (QUADRANT III)
N11 M23           (MIRROR MODE OFF)
N12 M22           (Y-AXIS MIRROR ON)
N13 M98 P8001     (QUADRANT IV)
N14 M23           (MIRROR MODE OFF)
N15 G28 X0 Y0 Z2.0 M09
N16 M30
%
```

Note the motion back to *X0Y0* at the end of subprogram. Missing this motion is the most common error when programming mirror images. Another block worth noting is the block N3 (main program). M23 — *mirror OFF mode* — is programmed for safety, in case of unexpected program interruption. This way, there is a guarantee that the program always start in the normal mode.

With rather minimal effort, programming mirror images can save significant amounts of programming time, when used properly. After all, imagining (or perhaps visualizing) the actual cutting motion provides the main key to success.

 Homeward Bound with G28
August 2006, updated February 2013

The mysterious G28 command. Its definition in a typical CNC machine manual is quite simple — G28 is a *machine zero return command*. However, that is where the simplicity ends. In my professional experience, there are two G-commands that can become a real headache for new programmers (and for a few seasoned ones as well). One relates to the usage of G41/G42: cutter

CNC Tips and Techniques

radius commands. As you guessed, the other one is the G28 command. In another essay (Cutter Radius Office — Basic Concepts), I touched upon the basics of G41/G42. Now is the time to demystify the other, often troublesome command: the mysterious G28.

Defining the G28 Command

In my view, the main cause of the difficulties lies in the much too common textbook-style definition (manuals, books). Understandably, command definitions have to be brief and often do not (or cannot) present the whole meaning of the command. Such is the situation with G28. If there were no limit to the definition of the G28 command, what would the definition be? Let me try this one:

G28 = Motion to machine zero applied to the specified axis (axes) and passing through an intermediate point

Although lengthier, this definition is quite a bit more specific, and it is the true definition of the G28 command. Keep in mind that G28, when used in the program, cannot be used on its own. It is just a non-modal preparatory command — a command that sets the stage, so to speak. G28 must always be associated with one or more of the available machine axes. Its purpose in the CNC program is to send the specified axis (or axes) to machine zero position, commonly known as the home position. On its own, the G28 is useless. It cannot be used without additional data, without at least one axis specified together with the G28.

Additional Axis Data

The block containing the G28 command must also include some other axis data; the letter X, Y, or Z without an assigned value is not good enough (at least on Fanuc and similar controls). So what should the value of the selected axis be? This was a

A Reader for Programmers

tough nut to crack during the initial software design, but the Fanuc engineers came up with a brilliant solution. They decided that the axis value (the axis data) associated with G28 will represent the intermediate tool location on the way to machine zero. Practically speaking, it means the CNC programmer can say, *"Yes, I want to move the tool to the home position, but I want to move it through a special point on the way."*

That point is called the *intermediate point*. It is a point to go through before the actual motion to machine zero is activated. One client of mine has compared this intermediate point with a stop at the Post Office on the way home from work. That is exactly what the intermediate point is all about — a position, a location, to go through on the way to machine zero.

Dimensioning

In a typical CNC program, the G28 command (with associated data) may be programmed in two modes of dimensioning: in the G90 absolute mode or in the G91 incremental mode. A typical program entry may look like this (Z-axis only shown):

G21 G90
...
G00 Z2.0
G28 Z2.0
M01

The same program segment in the incremental mode (G91) will be only a slightly different:

G21 G90
...
G00 Z2.0
G91 G28 Z0
M01

Both versions represent the same machining situation as

CNC Tips and Techniques

well as the method of machine zero return. Just remember the CNC basics: Z0 in the absolute mode (G90) is the part zero for the Z-axis whereas Z0 in the incremental mode (G91) means no axis motion along Z. When applied to the G28 machine return command, the absolute mode represents the intermediate point location (based on Z0), whereas the Z0 in the incremental mode represents zero motion (no motion) to the intermediate point of the G28 command. Many programmers like to change to the incremental mode because it is easier. There is nothing wrong with this method, as long as you remember to switch back to the absolute mode, when necessary.

Other Commands

It may also be worth mentioning three other commands available on Fanuc controls, all related to the original G28, machine zero return, command. They are G27, G29, and G30. In modern CNC programming, there is no need to use (or even understand) the G27 and G29 commands. Both are supported by the modern controls mainly for compatibility with older control systems. On the other hand, the G30 command (secondary machine zero return command) is very much alive and is defined as the secondary machine zero *return*. The G30 command uses the same principles as the G28 command, but is only available on those CNC machines that require it. For example, a CNC machine with a pallet changer will need two machine zeros.

Going home should always be a pleasant journey.

A Reader for Programmers

 Block Skip Adds Flexibility
September 2006, updated February 2013

From all the programming commands and functions, two are closely related to the settings at the operation panel of the CNC machine. One is the *Optional Stop*, using the M01 miscellaneous function. The other one is the *Block Skip* function (sometimes identified as Block Delete). The *Optional Stop* is quite straightforward: active (ON) or inactive (OFF). The *Block Skip* function is more complex and requires special care in the hands of both CNC programmer and CNC operator.

Block Delete

First, let's demystify the *Block Delete* description: no blocks will ever be deleted from the program. The term *delete* is misleading in this case. Blocks preceded by a forward slash will only be *skipped (bypassed)* if the operator chooses. In order to activate or deactivate this feature, the programmer must include at least one block in the program using the forward slash symbol and the operator must turn on or off the *Block Skip* switch on the machine operation panel. One without the other will not work. Because most programs do not require any block skip functions, the machine switch should be turned off as a general rule.

Block Skip

What is the purpose of the *Block Skip* function? Why would a block or blocks be programmed in the first place and then skipped on demand? There are several common uses (and many uncommon ones) for this feature. One of the most common applications is to optionally return one or more axes to machine zero, either at the beginning or at the end of a program. For example:

CNC Tips and Techniques

```
...
N34 G00 Z2.0
N35 G28 G91 Z0
/ N36 G28 X0 Y0
N37 M01
```

In this example, the Z-axis will always return to machine zero, whereas the return of X and Y axes is optional. This method allows faster cycle time for all parts, except perhaps the last one, when all axes must be at machine zero.

Another application for a block skip is to machine two parts from a single program, particularly if the differences are very small. The following example drills either four holes in a square pattern *(Part A)* or only three holes *(Part B)* with the third hole missing. The selection is done at the machine. Although it does not matter which part will be machined first, it is better to machine *Part B* first. Then if an extra hole is drilled by accident, it will become *Part A*, still to be machined.

```
...
N51 G99 G81 X5.0 Y3.0 R0.1 Z-1.74 F12.0
N52 X8.0
/ N53 Y7.0
N54 X5.0 Y7.0
N55 G80 ...
```

Note the repetition of Y7.0 in block N54. If the Y-location is missing, the machined pattern will be incorrect. This is a just a small example of how a simple oversight can lead towards a scrap. Other oversights range from relatively minor ones, such as a missing coolant function, to rather severe errors, for example, when a G01 cutting mode is skipped and G00 rapid mode takes over. In cases such as this, the result is not just a minor inconvenience, but a potential source for severe damage. A simple habit can eliminate all problems. *Always check the program flow twice —* once throughout, then by skipping the blocks with a slash. The

A Reader for Programmers

program must work in both cases to your satisfaction.

Example: Two Face Cuts

There are other applications of a block skip, such as programming a test cut for precise offset settings or programming two or more face cuts for a part with variable stock height or length. The next example shows two face cuts — the first one may be skipped if there is no need for it:

```
...
N78 G90 G00 X-3.0 Y2.65
/ N79 Z0.1 M08
/ N80 G01 X8.5 F15.0
/ N81 G00 Z0.2
/ N82 X-3.0
N83 Z0 M08
N84 G01 X8.5 F12.0
N85 G00 Z1.0
...
```

Although the M08 function could have been programmed earlier than in block N79, the program segment as shown serves as an example of necessary repetition in block N83.

Using the Forward Slash

Several general rules apply to using the forward slash in the program. Normally, the slash symbol must always be the first character in the block. Some controls allow block skip in the middle of a block, a feature that offers additional possibilities. This is not a standard feature, so check it first. Another rule is aimed at the CNC operators. Be sure to activate the switch *before* the first slash in the program is processed, and deactivate the switch only *after* the last block with the slash had been processed. Following these simple rules will make the *Block Skip* command a valuable addition to the current programming methods by adding flexibility at the machine.

CNC Tips and Techniques

Simulating the Toolpath
October 2006, updated February 2013

Any skilled CNC programmer and operator should be able to interpret and check the part program in its written form. Although time and experience will make such interpretation faster and more accurate, scrolling through a rather dry program code cannot beat a real visual simulation of the toolpath on a computer screen.

Types of Toolpath Simulation

Currently, there are several types of toolpath simulation available. One is built right into the control system, by the manufacturer, typically as an option at an added cost. Another type is an integral part of CNC oriented software, such as Mastercam, Edgecam, and most others.

For those machine shops that have only standard controls and no programming software, there is an attractive *stand-alone* alternative — toolpath simulation software that runs on a desktop or a laptop computer. Two common types of this group are 1) a wireframe simulation and 2) a simulation using a shaded model. Current software market offers a number of choices that vary greatly in features (and prices). Virtually all software vendors offer a time-limited trial period, which lets interested users test several software packages before purchasing the right one for their needs. Although the initial cost of the software is always important, its actual features and capabilities are much more critical, particularly in the long term.

All simulation software available today is expected to support the common tool motions, fixed cycles for machining holes, and several other common control features. Most have a built-in text editor that can be used to write and edit the part program. Unfortunately, a lot of software programs of this type lack many control system features and attention to detail. When selecting CNC simulation software, ask questions important to you and to

A Reader for Programmers

your common work, before committing to a purchase. Does the software support milling and turning operations in one package? Does it support special functions of the control system? For example, you may have *Coordinate Rotation* or *Programmable Mirror Image* functions that you use frequently, but cannot simulate them because the software does not provide the necessary support. Fanuc control support is generally standard for most simulation software, but does the software provide its variations, such as for Haas or Mitsubishi controls? What about *User Macros? Vertical lathes? DNC?* The list of these and similar questions can be quite long and it would take more than a short essay to cover all details. To offer at least some insight in the search of the perfect simulator, I will use my own experience as an example.

What I Want in a Simulator

My quest has been for a CNC toolpath simulator that I would want to include in one of my books (on a CD). I have evaluated over a dozen CNC simulation software packages. My goal was to find one that is user friendly, features rich, and affordable — a combination rather rare these days. Simulators using a shaded model were visually attractive, but the additional requirement of stock and tool definitions did not suit my needs for general use. Furthermore, I often did not find support of the software for many of the special functions that my in-depth CNC books cover, features I wanted to be simulated on the screen. I wanted simplicity, accuracy, and features — *a lot of features*.

Eventually, my long quest ended when I found (www.ncplot.com) a Michigan-based company offering software called *NCPlot*. Its version 2.0 was a relative newcomer to the market; the newest version is *NCPlot* v2.

What I like about *NCPlot* is what I also believe many machine shops will find very useful as well. The complete list can be quite long, so I will mention some of the main features that can be used as a general guide for selection of any CNC simulation software currently on the market. In NCPlot, once you

CNC Tips and Techniques

select the machine type (there are many customizable modes), the software interface is easy to navigate and selections are clearly defined.

NCPlot supports CNC mills and machining centers, including 4-axis machining centers. It also supports CNC lathes, including vertical lathes (right and left orientation). It even supports special variations between controls types, such as different subprogram formats. You can write the part program from scratch, using the full-featured built-in text editor (amazingly, it even renumbers Ps and Qs in lathe cycles), or you can import a DXF file from a CAD software to generate a toolpath.

Support for all basic tool motions, fixed cycles for machining holes, and clean graphic display with zooming, rotation, and panning is to be expected as bare minimum for this kind of software, but *NCPlot* goes much further. Its support for control options such as *Polar Coordinates, Scaling, Coordinate Rotation, subprograms* in many formats, G71-G73 type lathe cycles, *automatic corner breaking, DXF* input and output, and a whole lot more is first class. Add to it a full support for Fanuc macros (including compatible controls), macro expression calculator, even *Visual Basic* scripting, and you have software from a relatively small company that can challenge the big guys any day. Congratulations to Scott Martinez of *NCPlot* — keep on the great work.

The best part of *NCPlot* is its expected price. Back in 2006, the original version, for all these features and many more superfeatures, was under $300 (including a forum support and discount pricing for multiple purchases). Several years later, the price for the most recent version remains at $299. The combination of of features and price of *NCPlot* v2 seems to be unbeatable. Every machine shop has its own needs and unique requirements. When it comes to simulation software, you may decide to use *NCPlot* as the software of choice, or as software for the purpose of shopping comparison. Either way, an intelligent selection will result in both a better working environment and CNC programs

A Reader for Programmers

that reach the machine in much better shape. Let the example of my own needs serve as a path to your particular requirements.

 ## Automatic Corner Breaking
November 2006, updated February 2013

In a typical lathe work, there are many situations where the cut from shoulder to diameter — or from diameter to shoulder — requires a corner break. Breaking a sharp corner is a very common practice for eliminating burrs, providing clearances, and offering easier handling and assembly. It also improves appearance of the part.

Size of Corner Breaks

Engineering drawings often specify that all sharp corners are to be broken, but do not always provide their size. It is up to the CNC programmer to make the selection, which may vary from 0.005 to 0.020 inches (0.125 to 0.500 mm) or even more for special applications. The required corner break is either a chamfer at 45 degree angle, or a blend radius (fillet). Many corner breaks apply to cuts between a shoulder and the adjacent diameter, where the cut takes a 90-degree turn in one axis at a time. Although the calculations of start and end points are not difficult, they can be time consuming if many corners are involved (shaft work with many diameters, for example).

Programming Corner Breaks

Fanuc and similar controls provide a feature that greatly simplifies programming of such corners — it is called *Automatic Corner Breaking*. To better understand the difference in programming approach, compare the following examples. The first one shows a single chamfer of 0.1 between shoulder and diameter, calculated for each endpoint (blocks N5 and N6 make the chamfer):

CNC Tips and Techniques

N1 T0100 (0.1 X 45 CHFR - MANUAL CHAMFER)
N2 G96 S300 M03
N3 G00 G42 X4.0 Z0.1 T0101 M08
N4 G01 Z-2.0 F0.01
N5 X5.8
N6 X6.0 Z-2.1
N7 Z-3.0
...

The second example shows the same program modified for automatic chamfering (block N5 makes the chamfer):

N1 T0100 (0.1 X 45 CHFR - AUTOMATIC CHAMFER)
N2 G96 S300 M03
N3 G00 G42 X4.0 Z0.1 T0101 M08
N4 G01 Z-2.0 F0.01
N5 X6.0 C-0.1 (SHARP CORNER)
N6 Z-3.0 (CONTINUES IN Z-NEGATIVE DIRECTION)
...

The C-command in block N5 is the *chamfering vector* indicating the actual chamfer size. Note that the X6.0 indicates the diameter to reach (sharp point), as per drawing — no start/end points are required. When the control encounters a block containing the chamfering vector, it will automatically shorten the toolpath length by the C-vector amount. The vector sign is important. A positive C-vector means the chamfer is cut into the positive direction; a negative C-vector means the chamfer is cut into the negative direction (applies to both X and Z axes).

Even more effective is using the automatic chamfering for a blend radius. A blend radius between shoulder and diameter (or vice versa) is programmed in a very similar way as the chamfer vector, using the R-vector. Although the R-vector is a radius, *it is used in the G01 mode!* As with the C-vector for chamfering, the R-

A Reader for Programmers

vector also requires a positive or negative direction. The example below shows the previous example modified to cut an R0.1 blend radius:

```
N1 T0100 (R0.1 FILLET - MANUAL RADIUS)
N2 G96 S300 M03
N3 G00 G42 X4.0 Z0.1 T0101 M08
N4 G01 Z-2.0 F0.01
N5 X5.8
N6 G03 X6.0 Z-2.1 R0.1
N7 G01 Z-3.0
...
```

Using the automatic corner radius, the program is much easier to write:

```
N1 T0100 (R0.1 FILLET - AUTOMATIC RADIUS)
N2 G96 S300 M03
N3 G00 G42 X4.0 Z0.1 T0101 M08
N4 G01 Z-2.0 F0.01
N5 X6.0 R-0.1 (SHARP CORNER)
N6 Z-3.0 (CONTINUES IN Z-NEGATIVE DIRECTION)
...
```

Both methods are based on the same principles and use the same rules:

- The chamfer must have a 45-degree angle and the radius must have a 90-degree sweep angle, in a single quadrant.
- The programmed amounts of corner vectors are always *per side,* not diameter.
- The cutting direction *before* a corner break must be perpendicular to the cutting direction *after* the corner break.
- The cutting direction following the corner break must continue along a single axis only. It must be at least as long as the corner amount — it cannot reverse.

CNC Tips and Techniques

- Both chamfering *and* blend radius corner breaking take place in G01 mode (linear interpolation mode).

In programming, only the known intersection between shoulder and diameter is required. These rules apply equally to turning and boring CNC lathe operations.

Older controls do not support the C-vector for chamfers, and use I and K vectors instead. In this case, the I-vector is used to program the corner direction into the X-axis, and the K-vector is used for the direction into the Z-axis.

Automatic corner breaking provides a very effective method of programming corner breaks on CNC lathes and offers fast and easy changes, if necessary, without any recalculations.

 Working in Planes
December 2006, updated February 2013

Programming a three-axis CNC machining center generally means selecting *XY* point locations at a given Z-depth. This is typical to machining holes and standard contouring or pocketing toolpaths. There are many machining jobs, however, where the location point is not in the XY plane and the depth is not determined by the Z-axis. Welcome to programming in planes.

Have you ever wondered why many CNC programmers place G17 preparatory command at the top of program? I have found that many do it because they have *seen it somewhere*. Many programmers do not truly understand the *mystery* of the G17 command. By a book definition, G17 selects the *XY* plane. Definitions do not provide an explanation and their interpretation may be flawed; good understanding is very important. Programming a rapid motion G00 or linear motion G01 using any combination of the three axes can be done at any time, without special considerations, as long as it is safe for rapid motion or necessary during a cut. Current plane selection is always irrel-

A Reader for Programmers

evant in rapid and linear motions. (Motion will take place as specified.) The situation is much different for programming that involves the following three machining modes:

- Circular motion G02 or G03
- Cutter radius offset G41 or G42
- Fixed cycles using the G81-G89 plus G73, G74, and G76 commands

Defining the Plane

Only in these three cases do the CNC programmers have to watch for the currently active plane. A plane is always defined by two primary axes. Circular motion is the most suitable for understanding the concept; the principles apply to cutter radius offset and fixed cycles equally. Planes in CNC applications follow the standard mathematical definitions. The easiest way to understand mathematical definitions is to write the letters XYZ twice in succession then split the six letters into three pairs:

XYZXYZ   ...   XY ZX YZ, which define

G17 as *XY* plane, G18 as *ZX* plane, and G19 as *YZ* plane

To illustrate the concept of planes, consider the following program segment for a simple 180-degree arc:

G01 X20.0 Y35.0 F200.0
G02 X56.0 R18.0 (ONLY ONE AXIS SPECIFIED)
G01 X75.0

There is only a single axis specified with the 18 mm radius. How does the control interpret this block? An arc always requires two axes. Is the missing axis *Y* or is it *Z*? The plane selection in the program command provides the answer. If G17 is active, the missing axis is *Y*; if G18 is active, the missing axis is *Z*. G19 is not applicable because it does not involve the *X*-axis.

CNC Tips and Techniques

That is one purpose of the plane selection — to determine the missing axis in G02/G03 arc motions. In G41/G42, the plane selection determines left or right tool position relative to the contouring direction. In fixed cycles, it determines which two axes define the hole location and which axis controls the depth (an angle head is required):

- G17 — contouring in XY plane, depth in Z-axis — top view
- G18 — contouring in ZX plane, depth in Y-axis — front view
- G19 — contouring in YZ plane, depth in X-axis — right side view

The definition of a mathematical plane follows established international standards. When viewing a plane, the first letter always defines the horizontal axis in positive direction, the second letter defines the vertical axis in positive direction. When viewing a vertical machining center machine, the XY and YZ planes match the mathematical definition exactly. That is not the case in G18, which is defined as ZX plane mathematically, but viewed as XZ plane on the machine. Keep in mind that it is the *mathematical* definition that counts, not the way the machine is viewed.

When G18 plane is active, the CNC programmer must take utmost care while programming an arc motion, a cutter radius offset, or a fixed cycle. In circular interpolation, what appears to be a clockwise motion to the operator is actually a counter-clockwise motion in the program (and vice versa). What appears to be a tool location on the left side of the contour is actually a contour on the right side (and vice versa). In both cases, the G-code used is based on the mathematical plane, not the plane as viewed. Many errors are caused by misunderstanding this G18 anomaly.

Plane selection in fixed cycles is programmed less frequently because of special setup requirements. In order to machine a hole in a plane other than XY, a special angle head has to be used.

A Reader for Programmers

In summary, keep in mind that G00 and G01 modes are independent of plane selection, and G17 and G19 match the machine planes. The only one plane to pay special attention is G18. Also keep in mind that any plane has to be viewed in such a way that the first letter of the mathematical plane indicates the horizontal axis positive direction.

Readings From 2007

A Reader for Programmers

 ## A Case for Polar Coordinates
January 2007, updated February 2013

In a typical CNC program, the absolute coordinates XY indicate a point location measured from a common point known as part zero or program zero. This is the basis of the rectangular coordinate system, where the X point location is measured along the X-axis, and the Y point location is measured along the Y-axis (in G17 plane). Although this system is quite suitable for most CNC programs, there are times when it can become tedious and unproductive, because trigonometric calculations take time. Any programmer who ever used rectangular coordinates to calculate bolt holes location will agree. Bolt hole patterns or any arrangement of holes along an arc benefit from another type of point definition, the polar coordinate method.

Using polar coordinates on non-Fanuc controls is very common, for the simple reason that they are always available. Unfortunately, Fanuc controls offer this feature as an option only, and even that is not available on many old controls. Granted, if the feature can be added to the control, it is not very expensive, but can be a hassle anyway.

Defining Polar Coordinates

For the purposes of CNC programming, polar coordinates can be described as the endpoint of a line (often called a radius) that starts from another point (often called a pole) with a given length and angle. A polar angle is measured counter-clockwise from zero degrees (East direction or 3 o'clock position) as positive. Active plane selection does influence the use of polar coordinates, but the standard XY plane (G17) is the most common.

The major benefit of polar coordinates in CNC programming is that they virtually eliminate all calculations associated with the rectangular coordinate method.

CNC Tips and Techniques

Preparatory Commands

There are two preparatory polar coordinate commands for Fanuc controls: G15 (polar coordinate mode cancel or OFF) and G16 (polar coordinate mode active or ON). When G16 is programmed, it requires two inputs: a radius (length) and an angle. Note that there is no provision for defining the pole location directly with G16. Fanuc controls assume that the pole coordinates are X0Y0, which is not always the case. Many programmers consider this lack of definition a software design flaw. On the other hand, a G52 statement (local coordinate system) can be used to temporarily relocate the XY zero, so it's not a serious problem.

Sample Program

As an example of using polar coordinates, a program for a standard 6-hole bolt circle will be used, with the following data: pole center at X200.0 Y150.0 and bolt circle diameter of 120.0 mm. Polar coordinates are frequently used in conjunction with fixed cycles, for example, spot drilling cycle G82. There is, however, a major difference in the meaning of the XY addresses in the cycle. In the polar coordinate system, the XY data are defined as the arc radius (X) and the hole position angle (Y):

```
N1 G21
N2 G15 G17 G40 G80 T01 M06
N3 G90 G54 G00 X200.0 Y150.0 S1200 M03 T02
N4 G52 X200.0 Y150.0  (XY ZERO AT CENTER OF BOLT
   CIRCLE)
N5 G43 Z10.0 H01 M08
N6 G16              (POLAR COORDINATES ON)
N7 G99 G82 X60.0 Y0 R2.0 Z-2.8 P200 F175.0
N8 Y60.0
N9 Y120.0
N10 Y180.0
N11 Y240.0
```

A Reader for Programmers

N12 Y300.0
N13 G15 G80 Z10.0 M09 (POLAR COORDINATES OFF)
N14 G52 X0 Y0   (RESET BACK TO WORK OFFSET G54)
N15 G28 Z10.0 M05
N16 M01

The programming itself is very simple. Although the X60.0 (radius) does not change in this case, the Y (angle) increases by the amount of angle between holes (360/6 = 60.0), starting at zero degrees (Y0). No special calculations are required; the program is easy to understand and easy to change, if necessary.

There are several important features used in the program that require explanation:

As a precautionary measure, G15 has been programmed at the program beginning (N2), to cancel the polar coordinate mode, if active.

The first major tool motion should be to the pole — the bolt hole circle center (N3).

For a bolt hole pattern where the X-radius does not change, only the Y-angle increment (N8 to N12) is programmed

Either absolute or incremental mode can be used while polar coordinates are in effect.

Programming the angle with 0.001 accuracy makes a better program (three decimal places for an angle, if needed, for example, when programming a seven hole bolt circle).

In the less common situations that would require you to program polar coordinates in a plane other than G17, the meaning of addresses changes:

G17 plane XY => X = Radius, Y = angle
G18 plane ZX => Z = Radius, X = angle
G19 plane YZ => Y = Radius, Z = angle

CNC Tips and Techniques

Other rules and conventions apply the same way as for G17. For non-Fanuc controls that allow a direct pole point specification, there are additional advantages, not mentioned here. If you have many programs requiring holes located on an arc, an investment of just a few hundred dollars for the polar coordinates option will justify the expense in a very short time.

The "Other" Work Offset
February 2007, updated February 2013

A great majority of CNC programmers and machine operators are familiar with the work offset G54 (or the whole set of six work offsets G54 to G59 and beyond). Work offset in the program is used for one purpose only — to establish the location of part zero (program zero), which provides easier programming and operation. The programming side is easy. All you have to do is to include at least one work offset in the program (usually G54), then program all coordinates from the selected program zero. It is the CNC operators who have to prepare all the actual settings at the machine, particularly during setup. Operators may also adjust offsets during the production run, if necessary.

Basics of Work Offset

Let's look briefly at some basics. Work offset is the distance along each axis, typically along the X and Y axes only, measured from machine zero to part zero. Although the Z-axis can be included as well, in most applications it is set to zero, and handled by another type of offset (the tool length offset, G43 H). The reason is that each tool has a different length. Tool length offsets work exactly like work offsets, but for the Z-axis only. A part program may also include the cutter radius offset, using the G41/G42 command with a D-offset. For our purposes here, any reference to G41/G42 will be brief. Instead, the emphasis will be on the work and tool length offsets only because they are more

A Reader for Programmers

directly related. In order to understand these two offset groups, it is important to understand how they are stored in the control.

Tool Length Offset

Work offset has its own data entry screen, and so does a tool length offset. On most control systems, the tool length offset registry is shared with the cutter radius offset. That means the program cannot use the same offset number for the H-address as it does for the D-address. Programmers typically add an arbitrary number to the D-offset, based on the tool number. For example, tool T06 may use G43 H06 or G41 D56. Higher level controls may have geometry and wear offsets separate from each other.

Regardless of which type of offset memory the control system has, the end results will always be the same. In a typical application, each tool will have one length offset that controls the Z-axis (cutting depth). If the depth for a particular tool needs adjustment or fine tuning, the operator changes the tool length offset for that tool only.

Changing Length Offsets

Now, let's look at a fairly common situation where all length offsets have to be changed by the same amount. This can happen, for instance, when supporting parallels in a vise are changed after the initial setup. The least efficient method is to adjust each tool length offset individually. A better way is to adjust them all at the same time by changing the Z-setting in the work offset, which is normally Z0.000. For example, if each offset needs adjustment by 0.5 mm in positive direction, the Z-setting will be Z0.500. The question is which Z-setting?

The G54 Command

Work offset using the G54 command is the most common for the simple reason that it is first on the screen, and also a default in many cases. If a particular program uses G54 only, the Z0.500 should be set in the G54 screen. What about the other

CNC Tips and Techniques

work offsets? If there is also work offset G55 to G59 in the program and the 0.5 mm change applies to all offsets used, then the Z-setting of all work offsets used in the program has to be changed to Z0.500. Of course, the same applies to any additional work offset in the range of G54 P1 to G54 P48. Theoretically, up to 54 work offsets (6 standard + 48 additional) may require the Z-setting change. Although it is extremely unlikely you would ever face such a situation, changing more than one offset is not necessary, providing the change is uniform for all offsets programmed. The question of which Z-setting to change must be answered again.

Common or External Offset

Fanuc and similar controls offer another offset. This one does not have any number; in fact, it it is not even programmable, at least not through normal methods. This other offset is called the common or external offset. It is usually located to the left of the G54 offset screen and is identified either as COM (common) or EXT (external). Older controls used the COM identification, which was later changed to EXT, to avoid misunderstanding with COM, meaning communication. If the operator enters Z0.500 in the external work offset, it will affect all tools and all work offsets with a single common setting.

Using this external offset feature does not replace any adjustment of individual offsets, whether related to work coordinate system (work offsets) or tool length offsets. Also, this other offset has no influence at all on cutter radius offset. If you have to make a global change for all offsets, the other offset could well be the one to use.

A Reader for Programmers

 ## Going Helical with Threads
March 2007, updated February 2013

CNC thread cutting is a very common operation in many machine shops. Single point threading and tapping on CNC lathes and tapping on CNC machining centers are the common methods of thread development. Machining centers offer yet another method: thread milling. This method offers one of the most effective and cost efficient options when threading is a significant part of the business.

Helical Interpolation

The main key to any type of thread milling is the availability of a helical interpolation feature in the control system. Unfortunately, many controls offer this very useful feature as an extra option. Helical interpolation is a combination of circular and linear interpolations, both standard features. In more technical terms, helical interpolation is a simultaneous circular motion in two axes and one linear axis. The key word here is simultaneous — the rotating cutter follows the circular path (usually XY), while it moves up or down in the third axis (usually Z). Full circular motion will produce one pitch helix along the Z-axis direction.

Thread Milling

Thread milling is one of a few machining operations that offers a very extensive list of advantages and benefits, when compared with other methods of thread machining, or any machining for that matter.

- External and internal threads can be cut with the same cutter.
- Right hand and left hand threads can be cut with the same cutter.
- Full 100% thread depth is possible with most thread milling cutters.
- Higher cutting feedrates reduce cycle time.
- Smoother threads minimize or eliminate burrs.

CNC Tips and Techniques

- Secondary operations are eliminated, with fewer tools used.
- Precision and control increase when threading blind holes.
- Accuracy and quality increase.
- Spindle rotation does not change; spindle does not reverse.
- No broken taps means that hard materials are easier to handle.
- Small chips eliminate jamming in the cutter flutes, leading to better chip control.
- Only about 1/5 of the power rating is required.
- Tool life is extended.
- Taper threads (pipe thread) can be machined.
- Tooling and inventory costs are reduced.
- Higher concentricity is gained when compared with tapping.

... and many more

Tools for Thread Milling

The only relative disadvantage of this method is that tools (threading hobs) used for thread milling have the thread form built in, including the pitch. If the job calls for 12 threads per inch, you have to use a cutter with the appropriate pitch. For 16 threads per inch, a different cutter will be necessary. In this respect, it is no different from taps, except they cost a lot less.

Tools used are called thread milling cutters or threading hobs and come in several styles. Made of carbide, for small holes they look similar to taps, but without the helix built-in. For larger holes, you have a choice of single row cutters or multi-tooth cutters (with or without indexable inserts). A double insert tool can cut twice as fast as a single insert cutter.

A Reader for Programmers

Programming and Thread Milling

Surprisingly, there is not much extra that programmers have to learn to make thread milling work. There are no special G-codes associated with helical interpolation; preparatory commands G02 and G03 are used in a similar fashion as in circular interpolation. The remaining axis (the third axis) defines the linear motion, as in G03 X.. Y.. Z.. I.. J.., for the standard XY plane. Similar to normal circular interpolation applications, instead of the arc vectors IJK, the direct radius address R can be used as well (for partial arcs), providing the control system supports it.

As helical interpolation is typically an optional feature of the control system, how do we know the control supports this feature? Probably the easiest method to find out if the control has the helical interpolation feature installed is to try one simple block of program in the MDI mode. Make sure the spindle is away from the machine zero in the X and Y axes and enter these blocks:

G21 G91 G03 X-40.0 Y0 Z-2.0 I-20.0 J0 ...
    or a similar statement

If the control accepts the whole command, you should see the actual motion by watching either the spindle motion or the position data on the control screen. If the helical interpolation option is not installed, the control will issue an alarm.

Many holes that were previously machined with taps can also be thread milled. Thread milling will always produce better accuracy than the best tap. Part program controls the thread pitch; when using the cutter radius offset, the thread diameter can be adjusted at the machine with very high precision. Threading depth is uniform for each thread as well as over a large number of parts machined. Chip clogging is eliminated, also improving the thread surface quality. The surface finish on the threads is smoother than the surface generated by tapping operations. If any of these benefits are required, particularly for

CNC Tips and Techniques

a large volume machining, thread milling is certainly the machining method to consider.

Once you have helical interpolation available on the control, thread milling is not the only operation that can be of benefit. Helical interpolation can also be used for a number of non-thread-milling operations, for example, ramping into the material if plunging is either not possible or practical. Going helical can certainly improve the bottom line.

G76: Two Formats, One Cycle
April 2007, updated February 2013

Fanuc offers several multiple repetitive cycles for CNC lathe controls, in the range of G70 to G76. Without a doubt, for rough turning and boring, the G71 is the most commonly used cycle, along with G70 for finishing. G72 does the same job if the machining is more vertical than horizontal. Other cycles are used less frequently, mainly for special applications, with the exception of G76. This venerable cycle is the king of threading. It shines in every way you look, although that was not always the case. Somewhere between Fanuc 10/11/15 and the newest control models, Fanuc has changed the way this cycle is programmed. This change has affected the other cycles as well, but not in the same profound way as it affected G76.

The Original G76

Let's go back to the original cycle — the one written as a single block in the program. Its format was a bit lengthy, but you could feel the power that was contained in that one block of program:

G76 X.. Z.. I.. K.. D.. A.. F..
(with a P.. option for some controls)

A Reader for Programmers

The individual addresses are pretty straightforward – X is the thread final diameter, Z is the final position for the thread length, I is equal to zero, unless you're cutting a tapered thread, K is the actual thread depth, D is the first depth of the thread, A is the thread angle, and F is the thread lead as feedrate. The P option is not available on all controls; its purpose is to control the actual cutting motions of the tool. As good as this cycle was, and still is, it lacks certain features that could make the life of a CNC operator a bit easier. Consider the quick changes that can be done at the machine, in a rush to improve a thread on the fly, so to speak. There are only two: 1) changing the D to control the number of passes and 2) changing the A to control the infeed angle. That's it. All other minor adjustments were not readily possible, unless you got into the system parameters, and that is a territory forbidden to many.

Improving G76

Fanuc realized the need to improve this very powerful cycle, and made it even more powerful by adding certain features that can now be set right at the machine control, rather than through various system parameters. The result was a two-block G76 — yes, two blocks! — applicable to controls such as Fanuc 0,16,18,21, and a number of others. This format offers much more than the two features that can be changed at the machine. Is there a price to pay for this added convenience? Yes, but a rather small one. The G76 has to be repeated in both blocks. In addition, the programmers have to understand the meaning of each data entry within the cycle. Here is the format for a two-block G76 threading cycle:

G76 P.. Q.. R..
G76 X.. Z.. R.. P.. Q.. F..

Note the repetition of P/Q/R in both blocks — they are the reason for having two blocks in the first place. Needless to say,

CNC Tips and Techniques

each of these three letters has a different meaning, depending on the block. The explanation of each address will be a bit longer, because that is where the additional (and real) power is hidden.

In the first block, the most misunderstood is the P-address. It uses six (!) digits, in three pairs. Pair 1 is the number of finishing passes, which can be programmed in the range of 01 to 99. Pair 2 is the number of thread leads for a gradual pullout from the thread. This is a very useful feature when programming a thread that ends in a solid material rather than in a recess. In the case of Pair 2, the range is from 0.1 to 9.9, written without the decimal points. Finally, Pair 3 is the old A-address — the thread angle. Keep in mind that for both types of G76 input, this entry is limited to only six entries: 00, 29, 30, 55, 60, and 80. That takes care of the rather long P-address.

Addresses Q and R are quite easy. Q is the minimum cutting depth per side (without a decimal point). R is the amount left as finishing allowance (yes, this time, a decimal point is allowed). Both Q and R are very handy to fine tune the cutting conditions for a best thread possible.

Now to the second block. Most of the original settings are here: X and Z have the same meaning as in the one-block G76 cycle; R has replaced the original I for a tapered thread (R0 for a straight thread can be omitted); P has replaced the original K for the actual thread depth (no decimal point this time); Q has replaced the D as the first thread depth; but at least the F for feedrate (always the thread lead) has remained the same.

Without a question, the two-block G76 threading cycle is a real gem, but it can only be used on controls that support it. Watch whether a decimal point can be used or not; there are some possibly unexpected changes. This cycle is a winner over all. It cannot handle absolutely every thread there is (that's why we still have G32), but if you're in the special threading business, you have methods of your own already.

A Reader for Programmers

 ## Multi-Start Threading
May 2007, updated February 2013

For most threading applications on a CNC lathe, a thread is programmed with a single start. Programmers do not even think about the thread as having one start, they just take it for granted. The majority of threads in this category are designed for fasteners, such as bolts and nuts. Threads are also used for the purpose of transferring a motion, rather than for fasteners. Although these threads can also have only a single start, there are many that require more than one start. These are called multi-start threads, and they require special attention in programming.

The Theory of Multi-Start Threading

In order to transfer a motion rapidly over a relatively long distance, two, three, four, or more starts multiply the speed of motion. For example, take a standard external thread with eight threads per inch (8 TPI). It takes eight revolutions to travel one inch. With a double start, it takes only four revolutions, and with four starts, only two revolutions are required. Of course, a coarser thread can be used for the same result, but at a price — the thread depth will increase dramatically. Keeping the thread depth constant while speeding up the motion is the major reason for using multi-start threads.

In programming, the often misused term pitch must be understood as the distance between two adjacent threads — nothing else. Another term, the lead of a thread, takes on a special importance. Lead is the amount of axial travel in one revolution of the thread. For single start threads, pitch and lead are the same, hence the misuse of terminology. It is important to understand that the feedrate per revolution for threading is always the lead, never the pitch.

Take a single start thread of 8 TPI. Its lead and pitch are both 0.125, so the feedrate is F0.125. If the drawing specifies the thread

CNC Tips and Techniques

with a double start, the pitch will remain unchanged (0.125). Tthe depth will also remain unchanged, but the lead will double to 0.25, so feedrate F0.25 is required in the program.

Synchronization

Another unique consideration for multi-start threads is the synchronization of all start points. The control system guarantees synchronization for each depth, but the programmer has to include a new start position — a Z-axis shift — in the program. It is important that the start position for each thread is in such a location, that when viewed from the thread end of a screw or nut, each start point on the circumference will be divided in equal angular increments. A double start threads will be 180 degrees apart, triple start threads 120 degrees, and quadruple starts 90 degrees.

Writing the Program

Once you understand the theory behind multi-start threads, you can write the program using G32, G92, or G76 commands (Fanuc control). Summing up the theory in a few formulas may also help:

Feedrate = Thread lead = # of starts / TPI = #of starts * Pitch
Z-axis shift = Pitch
Number of shifts = Number of starts – 1

In another column (G76: Two Formats — One Cycle), I described in detail how the multiple repetitive threading cycle G76 works. To use this cycle for multi-start threading, I will use an external thread on a 3-inch diameter, 12 TPI and four starts. First the program (threading only, Z0 at the front face):

A Reader for Programmers

```
...
N21 T0600
N22 G97 S800 M03
N23 G00 X3.3 Z1.0 T0606 M08   (START POSITION)
N24 G76 P011060 Q004 R0.002
N25 G76 X2.8978 Z-2.5 R0 P0511 Q0120 F0.333333
N26 G00 W0.0833    (SHIFT FOR START 2)
N27 G76 P011060 Q004 R0.002
N28 G76 X2.8978 Z-2.5 R0 P0511 Q0120
N29 G00 W0.0833    (SHIFT FOR START 3)
N30 G76 P011060 Q004 R0.002
N31 G76 X2.8978 Z-2.5 R0 P0511 Q0120
N32 G00 W0.0833            (SHIFT FOR START 4)
N33 G76 P011060 Q004 R0.002
N34 G76 X2.8978 Z-2.5 R0 P0511 Q0120
N35 G00 X10.0 Z5.0 T0600
N36 M01
```

A couple of points are worth mentioning. One is the initial start position. Any thread requires about three times the lead to allow for feedrate acceleration to take place in the air. For a four-start thread in the example, this distance is rather large: one inch (N23). Each shift is programmed as an incremental movement of the pitch, even further away from the front face. Absolute positions can also be used, but they are more susceptible to miscalculations. Also note that each cycle repetition is exactly the same (feedrate is modal and six-digit precision is allowed for imperial threads). If you make a change in one cycle, you have to repeat the change in all others. The multi-start thread cutting application in the example is a good illustration of the programming concept, but does not take into consideration the effects of tool wear and other factors on thread tolerances.

CNC Tips and Techniques

 ## Automatic Tool Change — ATC
June 2007, updated February 2013

When it comes to tool selection for a particular job, the main focus of many CNC programmers is to select the most suitable diameter for the job. At the same time, the type of tool material, such as HSS or carbide, is also given serious consideration, as well as the number of flutes or cutting inserts. There are other issues, equally important, that should be considered when selecting cutting tools. They relate to the tool maximum diameter, maximum length, and maximum weight, as specified by the machine builder. Our consideration here is the type of ATC (Automatic Tool Changer) a particular CNC mill or machining center supports.

Fixed-Assess ATC

Manufacturers of CNC mills and machining centers provide two types of ATC: fixed-access type and random-access type. In the fixed-access type, the tool number used in the program must match the pocket number of the tool magazine. At the machine, it means the tool returns to the magazine pocket it came from, before the magazine is ready for the next programmed tool. This method is slow and quite inefficient, particularly in a high-volume production environment, although it provides a lower cost alternative in a non-production shop. Programmers are aware of this design feature and try to improve cycle times by arranging the order of tools more efficiently in the magazine. In spite of even the best efforts, the non-productive tool change can still be quite long.

Random-Access ATC

Most CNC machining centers offer much more advanced random-access tool changers as a standard feature. Their purpose is to minimize the tool change time, as it is impossible to eliminate it entirely. In this case, CNC programmers assign num-

A Reader for Programmers

bers to all tools without thinking how they will be distributed in the magazine. Tool numbers in the program do not have to match the magazine pocket numbers. One major difference in the program relates to the tool address T. While the letter T identifies the address as a tool number, it should really identify it as the ready tool number. The difference is subtle, but important. When the T-address is processed by the control, it will position that tool to a special standby or ready position of the tool magazine. This position is physically aligned with the automatic tool changer. Let's look at an example of a typical program start for the tool T01:

N1 G20
N2 G17 G40 G80 T01 (GET T01 READY)
N3 M06 (ACTUAL TOOL CHANGE)
N4 G90 G54 G00 X.. Y.. S.. M03 T02 (GET T02 READY)
N5 G43 Z.. H.. M08
... <MACHINING WITH T01>
N35 M01 (END OF T01)

Block N2 calls T01. At this command, the tool magazine starts rotating until T01 is located at the ready position. The tool change function M06 in block N3 will force the actual tool change between the current tool in the spindle (even if there is no physical tool) and the tool that is ready (waiting). This simple activity requires one very important step during tool setup. Because the CNC operator can place any tool to any magazine pocket (within reason), this additional setup step requires that each tool is registered in the control system memory. For example, T01 may be placed in magazine pocket number six and registered as such during initial setup; the control system will take care of the rest for as many parts as required. Even a non-existent tool in the spindle must be registered, usually as T99, for continuity. From this point on, no tool should be handled manually. Use the manual data input (MDI), if necessary, and pay attention

CNC Tips and Techniques

to the empty tool.

Some programmers prefer to combine the tool number with the tool change function in one block, for example, T01 M06. This is usually a matter of personal choice. There will be no difference in the actual tool change process or overall cycle time. My own preference is shown in the example. I prefer to split the tool number call (T01) and the tool change function call (M06) into two blocks. In automatic operation, there will not be any noticeable difference. During setup, when the operator often uses single block mode, this method provides time to check the tool in the magazine, before actual tool change.

Now to the example block N4. This block represents a typical first motion of any tool. It sets the control system to absolute mode, selects work offset, and provides a rapid motion to the first part location. It is also a block where spindle speed and spindle rotation can be programmed. Adding the next tool call in the same block makes sense. It takes place after the tool change, so the tool is ready for the next tool change, even if the cycle time of the current tool is short.

One other possible improvement in the area of tool programming is to repeat the next tool in the first block of the next tool, for example, the block N35 above will be followed by:

...
N35 M01 (END OF T01)

N36 T02 (GET T02 READY – IF NECESSARY)
N37 M06 (ACTUAL TOOL CHANGE)
N38 G90 G00 G54 ... T03 (GET T03 READY)
...
... <MACHINING WITH T02>

This method has one benefit — any tool programmed can be called at any time, generally when a problem is encountered that requires a tool repetition. Some programmers use other methods, for example, special block numbers.

A Reader for Programmers

 ## Maximum Tool Specifications
July 2007, updated February 2013

The focus of another column was to look at the types of automatic tool changers (ATC) and some related activities. Here, we look at other items also related to tooling in general and to each other in particular.

Maximum Tool Diameter

When selecting a tool diameter, its number of flutes or inserts is important for the actual cutting, but has no direct bearing on the tool change itself. However, the actual tool diameter does make a significant difference during tool change. Machining centers use a tool magazine that stores a specific number of tools in round pockets, adjacent to each other, typically in a round, oval, track, zigzag, or other tool magazine configuration. This configuration results in a certain distance — called a pitch — between pockets. This distance is always known from machine specifications, typically in the form of the maximum tool diameter that is allowed to be used. Maintaining a proper diameter is a serious safety issue; failure to adhere to the manufacturer's recommendations and specifications can have severe consequences. Machine manufacturers, being aware of the possible problems that machine shops often face, do allow an oversize tool diameter to be used in certain circumstances.

There is no provision in CNC programming that relates to maximum tool diameter. It is yet another important setup consideration, related to the CNC operator. The only effort a programmer can and should make is to inform the operator that one or more tools have an oversize diameter. This knowledge is absolutely critical during setup, particularly for safety reasons. Machine specifications typically allow a somewhat larger tool diameter, providing that magazine pockets on each side of such a tool are kept empty at all times. This situation must be achieved not only physically at the magazine, but in the control

CNC Tips and Techniques

system as well, during the tool registration process. Programmers should always know the maximum tool diameter allowed, as well as the extended maximum diameter with empty adjacent pockets. Using an oversized tool diameter is mainly an issue confined to the part setup at the machine.

Maximum Tool Length

By definition, machine manufacturers define maximum tool length as the actual tool extension from spindle face (gage line) to the tool tip. Such a distance is based on the actual machine dimensions and does not take into consideration any height of part or fixture located on the machine table. Again, for the reasons of safety and smooth operation, this distance should be known to the programmers, as well as the approximate height of the part setup. If the tool exceeds the maximum length only slightly, the programmers may decide to make the tool change in a particular safe position of the machine table, free of the part setup. Of course, the operator should be aware of this change from the normal, in order to keep this area clear. Additional test measurement at the machine is also recommended. Beyond the steps mentioned, there is not much programmers can do. As with an oversized tool diameter, it is mainly the operator's responsibility to provide safe machining conditions.

Maximum Tool Weight

Another machine specification provided by the manufacturer relates to the weight of the cutting tool. This weight relates to the automatic tool change, so it must always cover the combined weight of the tool holder, the pull-stud, the cutting tool itself, and any inserts and small hardware. Unlike with the maximum tool length, the program has no allowance for using a tool that exceeds this maximum weight only slightly. In fact, machine specifications do not include any heavier tool options at all. Well, perhaps there is one option to consider. Keep in mind that the maximum weight is specified for ATC only.

A Reader for Programmers

In spite of the availability of automatic tool change, programmers can choose to program the tool change conventionally — yes, changing it manually. The subject is too complex to squeeze into a short column. However, the actual programming process is not overly difficult, but requires some programming experience and a good attention to detail. It also requires good judgment because a heavier tool in the spindle will influence maximum spindle speed.

Changing the tool manually means the absence of both tool function T and ATC function M06 in the program. Instead, program end function M00 is used, with a suitable comment attached. CNC operators will place such a slightly heavier tool into the spindle manually, removing the current tool. The heavier tool will do all its work, and the program stops again. Now, the operators will reverse the process by removing the heavier tool and replacing it with the original tool. The program will continue normally from this point on.

Assuming the part program is itself structured properly, it is important to realize that working with oversized tools is not recommended as a standard practice, but as an exception only. When oversize tools are used, they should be treated very carefully, by increased vigilance during programming and especially during setup. In one way or another, they do add precious seconds or even minutes to the cycle time and have safety related implications.

 ## Control Features — Optional or Standard?
August 2007, updated February 2013

All CNC machines are made of two major parts: the machine proper and its control system (CNC unit). Although several machine builders offer their own controls (Okuma and Fadal come to mind), most manufacturers opt for a control independently made by Fanuc, Mitsubishi, Yasnac, or Siemens, to

CNC Tips and Techniques

name just a few. In all cases, the control system is designed to support as many machines and machine features as possible. A well-designed, generic-type control system can be easily modified to support certain unique machine-related features and accessories, such as part catchers, pallets, rotary axes, rigid tapping, and special thread cutting. Control support is necessary for these accessories to function properly; proper control support is always included as a standard feature. On the other hand, what about features that are related to operations rather than the machine?

Helical Interpolation

A good example is the helical interpolation feature. Its main application may be thread milling in many machine shops, but it can also be used for helical plunging into tough material (usually in pocketing) and for a variety of other purposes. Unfortunately, helical interpolation is not always a standard feature of the control system. A valid argument is, "Why should I pay for something I never use?", Whether this feature should or should not be standard on all controls is for debate.

Polar Coordinates

The whole picture is much clearer for another control feature (also a common special option on many controls) — polar coordinates. In my opinion, this very handy feature should be available on all controls, regardless of make, as a standard feature, never as an option. Polar coordinates beat rectangular system more often than not and are useful not just for machining bolt circles, but also for any holes located on an arc, even linear patterns of holes at a given angle. The best part is the holes do not even have to be equally spaced and no special fixed cycles are required. Although polar coordinates are very useful for machining holes, what about rotating a complete toolpath itself? It is relatively easy to program a rectangular pocket in orthogonal orientation, but try to program the same pocket at an angle,

A Reader for Programmers

even at different angles. Such a process can be time consuming and prone to errors.

Coordinate Rotation

Yet, the goal can be very easily achieved with, yes, another control option — coordinate rotation. A toolpath of virtually any shape can be stored as a subprogram. Then the subprogram can be rotated around a point by specified angle as many times as required using the coordinate rotation feature.

This trio of control features — helical interpolation, polar coordinates, and coordinate rotation — does not require any custom macros (which is another option). Each can be purchased individually. They are relatively inexpensive, which raises the question of why they cannot be part of the standard features lineup. Although it would be desirable to have all features available on every control, it would also increase the overall costs of CNC machine, and in many cases, unnecessarily. There are options that should remain options; if you need them, have the vendor turn them on for a nominal cost. Let's have a look at another trio, this time options that probably should remain options.

Other Options

The first of the three is a scaling function. It allows an existing toolpath to be machined larger or smaller by a specified scaling factor. One common application is the so called built-in shrinkage factor, either in flat machining or 3D machining. Another common control system option is direct drawing dimension programming. Although available for quite some time, it has really never reached its potential among users; being an option has to do a lot with it. The concept is quite straightforward — to simplify manual programming by minimizing calculations done by hand. Angles, chamfers, radiuses, etc., can be input as direct parametric values; then the control system calculates the required contour endpoints. This option may be worth looking into, particularly if you still do a lot of manual program-

CNC Tips and Techniques

ming. Last but not least is the already mentioned custom macro. In rather basic terms, a macro is a subprogram that uses variable input data, rather than fixed data. Although this feature offers tremendous opportunities and possibilities in programming, its utilization requires additional programming skills and the right type of machining to harvest its many benefits.

There are many other optional features that various control manufacturers offer to their customers. Take a look at them and see which ones you would like to become standard features. Is it reasonable to expect control manufacturers and vendors to include every current option as a standard feature? Not necessarily, but I strongly believe that at least the three options mentioned above should be provided as standard. After all, cutter radius offset used to be an optional feature as well at one time. Today, it's hard to imagine a control that would not include this feature in the overall price. Strangely enough, it is still listed as an option in many promotional brochures.

Fixed Cycles Repetition
September 2007, updated February 2013

Using fixed cycles for many holes is one of the oldest programming shortcuts available. Cycles G81–G89, as well as G73, G74, and G76 are main staples of everyday programming. Every CNC programmer knows that common data can be programmed only once and only new data has to be entered for each hole — one hole equals one block of the program.

Spot Drilling

Here is a sample program to spot drill a pattern of six holes:

```
N1 G20 G17 G40 G80 T01 M06
N2 G90 G00 G54 X1.0 Y1.25 S900 M03 T02
N3 G43 H01 Z1.0 M08
```

A Reader for Programmers

```
N4 G99 G82 R0.1 Z-0.14 P200 F5.0 (H1)
N5 X2.0 (H2)
N6 X3.0 (H3)
N7 Y2.5 (H4)
N8 X2.0 (H5)
N9 X1.0 (H6)
N9 G80 G28 Z1.0 M05
N10 M01
```

Spot drilling does not normally complete a hole, and there will be another tool or two applied to the same hole pattern. A classic application is spot drill, drill, and tap — three tools used for the same six holes. Although six holes do not present a problem, what about sixty or even six hundred holes? Is there a way to make further reductions in program length?

Reducing Program Length

In fact, there are two possible reductions available. Let's look at the first one. This method may not always be usable, as it depends on the actual hole pattern itself. Is there any equal spacing between the holes? If the answer is yes, this shortcut can save a few — or a lot — of blocks. In the example, there is a spacing of 1 inch between holes along the X-axis. Using the repetitive function K (or L on some controls), the program can be changed to incremental mode and shortened:

```
N1 G20 G17 G40 G80 T01 M06
N2 G90 G00 G54 X1.0 Y1.25 S900 M03 T02
N3 G43 H01 Z1.0 M08
N4 G99 G82 R0.1 Z-0.14 P200 F5.0 (H1)
N5 G91 X1.0 K2 (H2 - H3)
N6 Y1.25 (H4)
N7 X-1.0 K2 (H5 - H6)
N8 G90 G80 G28 Z1.0 M05
N9 M01
```

CNC Tips and Techniques

Six holes do not show the true picture, just the general idea. Although only one tool is used, the other two tools would be virtually identical in structure — just a different cycle, speeds, and feeds.

The second method is always available, regardless of the hole pattern. It can be combined with the first method, for an even shorter program. It involves writing a subprogram. A good programming approach is to include all six holes in the subprogram, not just the remaining five:

O7000
N101 G90 X1.0 Y1.25 (H1)
N102 G91 X1.0 K2 (H2 - H3)
N103 Y1.25 (H4)
N104 X-1.0 K2 (H5 - H6)
N105 M99

How does this method change the structure of the main program? Block N4 will machine the first hole, regardless of which cycle is used. Because the first hole is included in the subprogram, it will be machined twice. This is not acceptable and some change in the program must be made. Fanuc and similar controls provide a very useful function for exactly this kind of situation: K0 or L0. Normally, the control system defaults to K1 or L1 — use it just once. Of course, K1 and L1 are not written in the program;, they are assumed. What K0 or L0 does is quite simple. They will store the cycle data into the system memory, but they will not cut the hole at that location. The final version of the program (one tool only) is shown next:

N1 G20 G17 G40 G80 T01 M06
N2 G90 G00 G54 X1.0 Y1.25 S900 M03 T02
N3 G43 H01 Z1.0 M08
N4 G99 G82 R0.1 Z-0.14 P200 F5.0 K0 (NO HOLE)
N5 M98 P7000 (H1 - H6)

A Reader for Programmers

N6 G90 G80 G28 Z1.0 M05
N7 M01

Remember to switch to incremental mode. Otherwise, all repetitions will take place at the same location. Depending on the number of holes, these two methods can save literally hundreds of program blocks, making the final program easier to interpret, change, and store. If used properly, K0 (L0) can be a very powerful tool.

Programming Process — When Is It Completed?

October 2007, updated February 2013

Regardless of which type of CNC programming is used in a machine shop — manual or computer assisted — the basic programming process is generally the same. Start with the drawing; evaluate material; choose tools and setup, cutting conditions, and toolpath; and write the program. When a CNC program is fully completed and released to the machine shop for production, it seems that the programming process is over and the part programmers usually move on to another project. After all, calculations have been done, decisions have been made, the program has been written, it is well documented, and the program file is on the way to the CNC machine. Is the programmers' job for such a part really finished? Is there some possibility that could bring the program back for some reason, perhaps with an operator's comments, suggestions, or even some constructive criticism?

The Programmer's Role

Of course, if the part program works perfectly, without any changes, the programmers will usually not hear a word from any direction. On the other hand, the programmers will hear nega-

CNC Tips and Techniques

tive comments and criticism from all directions. So, the question is: When is the programmer's responsibility for a particular part really over? At what point in the manufacturing process can the programming results be evaluated? At what point can the program qualify as a very good (or even perfect) program?

The most likely answer, and probably the fairest and most reasonable answer, would be whenever the part has been machined under the most optimized working conditions. This means that the programming responsibility does not end with the initial program and documentation delivery to the shop. The part program at this stage is still very much in the development process; it is unproven. It still has to be loaded to the CNC system. The machine has to be set up, cutting tools mounted and measured, and a variety of small jobs completed before the first part can be started. Then, the program has to be proven, possibly optimized. True, all these tasks are the sole responsibility of the CNC operators, so there is no need for the programmers to care what happens during machining, right? Not really.

Working with Operators

CNC operators can only work with a program as provided; they can be expected to make only minor optimization changes, if necessary. All CNC programmers should make an effort to be in constant touch with the actual production team. A good comparison is with the field of complex business software development. In this environment, it is necessary to have a team of specialists working on a certain large programming project, where all individual parts have to fit together flawlessly. After all, most programming ideas are the result of talking to colleagues and the actual users of the program or particular software.

The same is true for CNC programs used in machine shops. The actual users of part programs are typically CNC machine operators. They are directly involved in the actual production, and they provide a very deep gold mine of constructive ideas, improvements, and suggestions. Talk to them, ask questions,

A Reader for Programmers

provide suggestions, and — most important — listen to what they have to say. Programmers who never put their foot in the machine shop or who go there reluctantly are on the wrong track. The same is true for programmers who go to the machine shop with their eyes closed and their ears plugged, and programmers who take the attitude that they are always right.

Exchanging ideas with CNC machine operators, asking questions, and seeking answers is the only way to be fully informed about what is actually going on in the machine shop. It is in the programmers' own interest to know how the CNC operators feel about their programs, the programming style, and their approach to CNC programming overall. Exchange thoughts and communicate with each other — that is the best way to become a better CNC programmer (as well as CNC operator). Most machine shops offer tremendous resources, take advantage of them.

These days, more than ever, CNC technology is an instrument to improve productivity with a minimal human involvement, measured at least by the physical level. As with any other technology, it must be understood and managed intelligently, by qualified people with experience. Without a firm grip and good control, without good management, and without the right attitudes, the technology will not yield the expected results; in fact, it will become counterproductive.

In order to develop high quality programs, CNC programmers must have certain qualities of their own. The quality of actual parts produced is measurable. Is the quality of CNC programming also measurable? We look at this subject in the essay *Quality in CNC Programming*.

CNC Tips and Techniques

Quality in CNC Programming
November 2007, updated February 2013

In the article Programming Processes — When Is It Completed?, we looked at improving the quality of the CNC programming process. Without qualified programmers, such a process is bound to fail. Quality and efficiency in a competitive market is the major focus of many manufacturing facilities. Whether in a large corporation with a workforce of thousands or in a one-person machine shop, quality should always be the top priority in all fields. In CNC, the focus is more often on the quality of the finished parts, where it should be. Quality of parts machined is the final outcome of many prior steps, including — in a large part — CNC programming.

All parts machined on CNC equipment are evaluated by their quality, usually during machining or after they are machined. Quality inspectors check many part features: Are the dimensions within tolerances? Is the surface finish up to standard? Is there a consistency between parts, etc.? Modern CNC equipment provides optional inspection features during the machining process, such as in-process gauging, using dedicated equipment, and special custom macros. Many machine shops, particularly the smaller ones, require their CNC machinists to be quality inspectors while the parts are being made. One subject that is equally important, yet not very often mentioned, is quality in CNC programming.

Qualities of a CNC Programmer

The proper approach to CNC programming should always start with a solid plan. Although number one quality of a CNC programmer is knowledge and skill, there are at least two related and equally important considerations in part program planning: the programmer's personal approach and the professional attitude. How CNC programmers approach a certain job, assign-

A Reader for Programmers

ment, or project will greatly influence the final outcome of all parts produced. Programmers' attitudes also have significant influence on program development and final results. They also have significant influence on CNC operators or setup persons, and, of course, on actual production.

Ask yourself some questions. As a CNC programmer,
- Are you attentive to detail?
- Are you precision minded?
- Are you well organized?
- Are you people oriented?
- Are you concerned if something is not done right?
- Do you cut corners just to have the job done?
- Do you guess instead of calculating?
- Can a CNC program you have just developed, or a program written earlier, be improved to make it safer and more efficient?
- Do you develop programs for portability, so they can be used on more than one machine?

Although many programs delivery top notch parts, it is virtually impossible to make a perfect program. As much as perfection is a noble goal, it is not the goal in itself.

Program Quality

CNC program quality is much more than writing an error-free program. That is the absolute prerequisite and goes without saying. Quality in CNC programming includes concern about how a particular program affects the CNC operators, overall machine setup, and the actual part machining. Quality in CNC programming requires a constant effort at improvement and a desire to make the next program even better.

Consistency in part programming is one the best ways to achieve high quality programs. Once a certain method or process has been found superior to others, stick with it. Use the same method again and again, at least until an even better method can be found. CNC operators like nothing less than programs that vary in structure. Consistent program structure is the first key to consistency. Structure in programming is more than just uniform formatting; It is a well thought out overall approach that considers not just the current part, but future parts as well. In order to develop a consistent program structure, it is important to consider all possibilities that may happen and ask a few "what if" questions.

Larger shops may have more than one person developing part programs. Although it is more difficult to reconcile various programming preferences and styles, some consensus is always necessary. CNC operators should not be forced to change their working methods just to suit each programmer's style.

Whether programming manually or using a CAM programming system, part complexity should never stand in the way of final quality. Quality should be built in to simple and complex programs equally. Quality is related only to your knowledge level, willingness to solve problems, and ability to listen to ideas. It should be a personal goal to make a program — every program — the best program possible.

Set your quality standards high!

Short Ideas and Observations
December 2007, updated February 2013

When I talk to the many CNC programmers and operators in companies large and small, it never ceases to amaze me their depth of knowledge. I have seen some ingenious ways of solving seemingly unsolvable problems. I have seen dedication and eagerness at its best and learned from many. I have also seen some serious gaps — a lack of understanding some basic programming issues. They are often small items that can have a big

A Reader for Programmers

impact. This is the subject of this article, in the form of a few short statements and a short explanation.

Work Offset G54 is NOT Always an Automatic Default on Most Controls

All six work offsets that most Fanuc and similar controls offer are seldom needed in a single program. That leaves only one that is needed, G54 (being the first in line). Many programmers do not use it at all, and have no problems. The reason is that they count on the default control settings. Some controls will use G54 settings, even if G54 does not appear in the program. Controls that do not have G54 set as a default may cause erroneous motions, including a possible over-travel. It is always a good practice to include work offset in the program and not count on default settings.

G92 and G54 are Quite Different

Now mostly obsolete, the old G92 position register command is always measured from part zero to the current tool position. The modern work offsets in the range of G54 to G59 are measured from machine zero to the part zero. Conventional wisdom says that these two groups should never been used in the same program together. I have seen a few successful exceptions, but the wisdom is still correct — use one or the other if you have a choice, but not both.

Selecting Offset Numbers H and D

There is a fairly common practice among programmers to match the tool length offset number (address H) to the tool number (address T). For example, programming T03 along with H03 makes sense, as only one tool length offset is normally required per tool. When it comes to the cutter radius offset, the D offset number can also be the same, providing the control has a separate register for each offset group. This feature is only available

CNC Tips and Techniques

on higher level controls or as a special option. Most controls have a shared offset register — both tool length and radius offsets are in the same memory storage area. In this case, the D-offset number should be different (not the H-offset) because it is not required for each tool. A typical approach is to change the D-offset number arbitrarily by a certain amount, usually by 20, 30, 40, or 50, depending on the actual number of tools available. If tool three requires both offsets, the program will use T03, H03, and D53, for example. Don't forget H.. is programmed with G43, and D.. is programmed with G41/G42.

Dimensional Units — Imperial and Metric

Common program selection for imperial or metric units is from two G-codes: G20 (imperial) and G21 (metric). In most cases, they replaced the old G70/G71 commands, common on early control systems. Although many programmers bundle either of the two G-codes with other G-codes in a single block at the beginning of a program, this may prove to be a poor practice. The reason is that some machines require having G20/G21 on a separate block. The reason is that some older controls do not convert offset settings from inches to millimeters or from millimeters to inches; they only shift the decimal place by one position. In this case, the G-codes can be bundled together. Newer controls do actually convert offset and other settings from one unit system to the other. To do that without any conflict, G20 or G21 should — or even must — be programmed by itself in a block. This approach will never cause harm on any control and makes such program more portable.

M-Functions Together with a Motion

Several miscellaneous functions should be programmed in a block without any other data. These include M00, M01, and M30. Others are programmed logically and tied to other functions, for example, spindle speed and tool rotation (S1000 M03). What about M-functions programmed together with a motion?

A Reader for Programmers

Take two examples: G01 X10.0 F15.0 M03 will cause the spindle rotation simultaneously with the motion, whereas G01 X10.0 F15.0 M05 will complete the motion first, and then turn the spindle off. In this respect, the control behaves quite logically.

... And Finally, Some General Observations

Alarm lights on the operation panel can burn out and fail to warn the CNC operator. Check them frequently.

The miscellaneous function M19 (or M20 on some controls) can be used to orient spindle through the program. Spindle can also be oriented at the machine or control using the MDI method or a physical button.

G91 G28 X0 Y0 Z0 (milling) or G28 U0 W0 (turning) are the fastest ways to send machine axes to their respective zero positions from any tool position, providing it is safe!

Readings From 2008

A Reader for Programmers

 ## Spindle Speed Control on CNC Lathes
January 2008, updated February 2013

Programming spindle speed on a CNC machining center requires nothing more than specification of the number of revolutions per minute and rotational direction. For example, 1500 r/min at normal spindle rotation will be entered as S1500 M03. Because the cutting tool diameter does not change, neither does the spindle speed. Although the principles are the same, programming spindle speed for CNC lathes has to consider several additional factors.

First, for turning and boring, it is the part that rotates, not the tool. Therefore, the part diameter changes frequently, sometimes quite dramatically, during machining. With every change in diameter, a corresponding spindle speed is required. Second, some machining operations take place at the spindle centerline, such as drilling, reaming, and tapping. Then, there is single point threading.

G96 and G97

Fanuc and similar controls provide two programming modes: cutting speed and spindle speed. Cutting speed is programmed in feet or meters per minute; spindle speed is programmed in revolutions per minute. To select one or the other, two G-codes are available: G96 selects cutting speed and G97 selects spindle speed. Cutting speed is known under several other descriptions, such as constant surface speed or peripheral speed. Abbreviations such as SFM or SFPM (surface feet per minute) are also used for imperial units. In both cases, either the G96 or G97 command is followed by the S-address. A bit of care is necessary here to avoid making an error. For example, in imperial units, the program blocks may be:

```
G96 S400 M03    ... selects 400 feet per minute
G97 S400 M03    ... selects 400 revolutions per minute
```

CNC Tips and Techniques

A single digit error can have serious consequences. When do we use one mode over the other?

Rather than listing when G96 is used, it is easier to remember when G97 is used. Spindle speed mode G97 is used for all operations that take place at the machine centerline. The already mentioned drilling G97 mode is also used for any threading operations, regardless of cutting diameter. That's it. All other operations will use G96, the cutting speed mode. Cutting speed will automatically calculate the actual spindle speed, at any diameter. The result is constantly changing spindle speed, as the tool moves from one diameter to another.

Just a couple of issues that belong to the caveat emptor category should also be addressed. One issue is what will happen if G96 is used in drilling, for example? Because the cutting diameter at machine centerline is zero, the common spindle speed formula fails. In this case, the spindle will rotate at the maximum speed of the selected gear range, which can be extremely high. A related issue is a facing operation that moves the cutting tool to the centerline or just slightly below. G96 mode is desired, but the proximity to a zero diameter at the end of cut may increase the spindle speed too high and even present a safety risk. In this case, there is a simple remedy. Command G50 can be used to limit the maximum spindle speed in G96 mode. Programming — for example, G50 S2500 — will gradually increase the spindle speed as diameter is decreasing, until the control reaches 2500 r/min. At that point, this pre-selected spindle speed will be clamped and remain that way until the end of the cut. There may be some increased tool pressure, which can be minimized by a slower cutting feedrate.

Potential Problem with Cutting Speed

On some machines, particularly older ones, there may be a small, but possibly significant problem with cutting speed that is often overlooked. Consider the following example, representing the first few blocks of a typical turning or boring operation:

A Reader for Programmers

N1 G20 T0100   ... select imperial units and tool 1
N2 G96 S450 M03   ... 450 feet per minute selected
N3 G00 G41 X0.7 Z0 T0101 M08   ... move tool to target diameter 0.7
N4 ...

Now, ask yourself a question: Can you precisely calculate the actual spindle speed, when block N2 is processed? The target diameter is 0.7 from the current tool position, correctly specified, but after the spindle command. When block N2 is processed, spindle speed will be calculated for the diameter stored in the geometry offset. This is the only diameter known to the control at this point. How important is this knowledge?

Consider the geometry offset for the X-axis (diameter) to be 23.5 inches. Spindle speed calculated at this diameter will be only 73 r/min for 450 ft/min. Spindle speed at block N2 will be 73 r/min. In block N3, the diameter has changed (rather abruptly) to 0.7, so the control will correctly recalculate the spindle speed as 2455 r/min. Such an extreme change may present a problem on some machines. Depending on how the control system handles spindle acceleration, the cutting tool may have a physical contact with the material before the speed of 2455 r/min has been fully reached. The solution to this possible situation is simple, but it does require manual calculation. In the example, spindle speed for the final diameter of 0.7 can be pre-programmed with the G97 command.

The standard example above will be modified by adding G97 S2455 M03 before the tool motion:

N1 G20 T0100   ... select imperial units and tool 1
N2 G97 S2455 M03   ... preset r/min based on target diameter
N3 G00 G41 X0.7 Z0 T0101 M08 ... move tool to target diameter 0.7
N4 G96 S450 M03   ... 450 ft/min for all subsequent cuts
N5 ...

A simple addition of the pre-calculated spindle speed solves the problem. Spindle speed expected at 0.7 diameter will

CNC Tips and Techniques

be in effect before the tool reaches the material.

Once you understand the differences between G96 and G97 commands, fine-tuning any lathe program with G50 or pre-calculated spindle speed should present no difficulties.

Live Tooling on CNC Lathes
February 2008, updated February 2013

In the early days of numerical control, two-axis lathes and three-axis machining centers were the main production machines of many shops. Quite a lot has changed since then and the CNC lathes today are much different than ever before.

Advances in Machining Technology

True, the standard two-axis lathe is still around for traditional work, and for general round and conical machining. This original concept has been maintained, but new features have been added to it. In the late 1980s, several machine tool manufacturers added a simple milling and drilling capability to their lathes, allowing a completion of simple parts in one setup. These additional tools had to work under their own power — they had to be power driven or live when the lathe main spindle was stationary. The term live tooling was born and collectively describes all non-turning tools. Machines using this technology are often called turn-mill machines.

Over the last 25 years, machining technology has advanced dramatically. It became possible to design CNC lathes capable of many very complex machining processes. Live tooling is the first step towards multi-process machining — combinations of various turning and milling applications, multiple turrets and chucks, sub-spindles, automated part reversal and transfer plus many other features.

A Reader for Programmers

For CNC programmers, knowledge of the various designs is very important, as programming will somewhat vary. Most common live tools used are tools to machine holes and end mills for milling operations. On slant bed type lathes, the live tools can be mounted parallel with, or perpendicular to, the machine centerline. That means machining can be completed on the part face or its circumference. The standard XZ motion is supplemented by a third motion: the C-axis. The C-axis controls spindle rotation in degrees when live tools are used. Normal spindle rotation in revolutions per minute is used for standard turning, boring, and related operations. The main purpose of the C-axis is to rotate the main spindle in direct relationship with the motions of live tools.

A number of miscellaneous functions (M-functions) are used with the C-axis and they are not always consistent between manufacturers. M-functions control rotation of the main spindle as well as rotation of the driven tool. A pair of M-functions controls whether the main spindle is ON or OFF. Although some machines may lock the C-axis automatically, others require another pair of M-functions. These functions and others are machine specific and manufacturers list them in their machine documentation.

A Major Restriction

Working with XZ+C axes has one major restriction. Drilling, reaming, tapping, simple slot machining, etc., can be completed successfully; however, the location of the holes or slots must be on the plane that intersects the machine centerline. Using slotting as an example, only a slot corresponding to the width of the end mill can be machined. When the C-axis is supplemented with another axis (the Y-axis), drilling and milling operations can be completed off machine centerline, greatly enhancing machining flexibility. Y-axis is factory installed in addition to the XZ+C axes. Its motion, which is perpendicular to both X and Z axes, is similar to a typical CNC vertical machining center.

CNC Tips and Techniques

Programming Methods

For holes machined with live tools, programming methods include special versions of standard fixed cycles for milling: one set for face holes and one set for holes located on the part circumference (diameter). A small example will drill two holes on a part face:

```
G20                      (IMPERIAL UNITS)
M14                      (SPINDLE MODE OFF = C-AXIS MODE ON)
T0404                    (TOOL SELECTION)
G97 S850 M103            (LIVE TOOL SPINDLE SPEED -
                           CW ROTATION)
G98 G00 X4.0 Z1.0        (FEED/MIN AND START POSITION)
C90.0                    (HOLE 1 ORIENTATION)
G83 Z-1.5 R-0.9 F10.0    (DRILL HOLE 1 - R IS INCREMENTAL !)
C120.0                   (DRILL HOLE 2)
G80 G99 X10.0 Z5.0 M105  (TOOL INDEX SAFE POSITION - LIVE
                           TOOL STOP)
M15                      (SPINDLE MODE ON = C-AXIS MODE OFF)
```

Interpolation Methods

In addition to fixed cycles, CNC lathes with live tooling also offer two interpolation modes that further increase the number of machining operations that can be completed in a single setup. One of these features is polar coordinate interpolation; the other is cylindrical interpolation.

Polar coordinate interpolation is a feature that automatically converts rectangular coordinates to polar coordinates on a continuous basis. Programming shapes such as square, hexagon, octagon, and other polygons, can be simplified by using polar coordinate interpolation. If this programming method is not available, continuous simultaneous machining of flats along X and C axes is possible, but very time consuming. Complex toolpaths will benefit from CAM software, but simple flats can be programmed manually. CNC machining centers offer a feature called Polar Coordinates using commands G15 and G16, not to

A Reader for Programmers

be confused with Polar Coordinate Interpolation.

In cylindrical interpolation, the C-axis can also be used together with the Z-axis to program a continuous contour toolpath. This method provides synchronized motion of the linear Z-axis with C-axis part rotation. The most common application for this type of machining is cylindrical grooving.

This brief overview touched upon only some of the manual programming methods that can be used with C and Y axes. For more complex parts (especially contours), using a suitable CAM software is much more efficient.

 Trial Cut for Measuring
March 2008, updated February 2013

Precision CNC machining is always a cooperative effort between CNC programmers and CNC machine operators. Programmers can employ certain program features that will enable the machine operators to perform functions that would otherwise be difficult or even impossible. One such method is programming a trial cut for measuring.

There are many factors in machining that influence the final size of a part, such as heat, tool deflection, tool wear, and even lubrication to some extent. For those parts that require very tight tolerances, the negative effect of these factors may result in an out of tolerance size and even a possible scrap. Another problem that is quite common is machining certain hard-to-measure shapes, such as a cone on CNC lathes. By using a simple programming method known as the trial cut, CNC programmers can greatly influence the final size of a part.

Measuring Cones

Take the already mentioned cone as an example. Without a suitable custom gauge, a cone is a difficult shape to measure at the machine, as there is no flat spot for the micrometer to use.

CNC Tips and Techniques

Even if such a special gauge is available, it will be too late to find out that the cone is out of tolerance. The method of using a trial cut is to provide a special flat area (diameter) that can be easily measured, before the final cone is machined, following a few simple rules.

Normally, a cone machined from round stock, such as the 5.0 inch diameter used in the following example, will include G71 roughing cycle and G70 finishing cycle in the part program:

N1 G20 T0100 (ROUGHING TOOL)
N2 G96 S300 M03 (CUTTING SPEED)
N3 G42 G00 X5.2 Z0.1 T0101 M08 (START POINT FOR CYCLE)
N4 G71 U0.15 R0.025
N5 G71 P6 Q8 U0.07 W0.02 F0.015
N6 G00 X2.8 (CONE START)
N7 G01 X4.6 Z-3.5 F0.008 (CONE END)
N8 U0.2 (RETRACT FROM PART)
N9 G40 G00 X10.0 Z4.0 T0100 (CLEAR POSITION)
N10 M01

N11 T0300 (FINISHING TOOL)
N12 G96 S500 M03 (CUTTING SPEED)
N13 G42 G00 X5.2 Z0.1 T0303 M08 (START POINT FOR CYCLE)
N14 G70 P6 Q8
N15 G40 G00 X10.0 Z4.0 T0300 (CLEAR POSITION)
N16 M30
%

Although correct, the program above does not offer any means to measure the conical part efficiently. By adding several blocks of trial cut in the program and using the finishing tool (T03), a suitable straight cut can be machined first.

When measured, various offsets can be adjusted as necessary — all this is done before any roughing cuts take place:

A Reader for Programmers

```
/ N1 G20 T0300 (FINISHING TOOL)
/ N2 G96 S500 M03 (CUTTING SPEED)
/ N3 G42 G00 X5.2 Z0.1 T0303 M08 (INITIAL POSITION)
/ N4 X4.925 (DIAMETER TO BE MEASURED)
/ N5 G01 Z-0.35 F0.008 (LENGTH OF CUT FOR MEASURING)
/ N6 U0.2 (RETRACT)
/ N7 G40 G00 X10.0 Z4.0 T0300 (CLEAR POSITION)
/ N8 M00 (MEASURE DIAMETER - MUST BE 4.925)

N9 G20 T0100 (ROUGHING TOOL)
N10 G96 S300 M03 (CUTTING SPEED)
N11 G42 G00 X5.2 Z0.1 T0101 M08 (START POINT FOR CYCLE)
N12 G71 U0.15 R0.025
N13 G71 P14 Q16 U0.07 W0.02 F0.015
N14 G00 X2.8 (CONE START)
N15 G01 X4.6 Z-3.5 F0.008 (CONE END)
N16 U0.2 (RETRACT FROM PART)
N17 G40 G00 X10.0 Z4.0 T0100 (CLEAR POSITION)
N18 M01

N19 T0300 (FINISHING TOOL)
N20 G96 S500 M03
N21 G42 G00 X5.2 Z0.1 T0303 M08 (START POINT FOR CYCLE)
N22 G70 P14 Q16
N23 G40 G00 X10.0 Z4.0 T0300 (CLEAR POSITION)
N24 M30
%
```

The first eight blocks are shown with a block skip function, represented by the / symbol. It is more likely that the actual measuring will not be required for each part. This approach to part programming is more efficient because it eliminates a re-cut and greatly enhances accuracy. Setting the block skip mode OFF, CNC operator checks the trial dimension (note the message attached to M00 program stop function), adjusts the suitable offset if necessary, and continues machining with block skip set ON. Note that the trial cut was made with the same tool that will be used for finishing.

CNC Tips and Techniques

When selecting trial cut diameter, make sure the tool removes about the same amount of material that is removed in the finishing cut. In the example, the trial cut diameter is 4.925, resulting in the actual depth of cut (5.0 − 4.925) / 2 = 0.0375 per side, which is almost equivalent to the amount of stock left for finishing: 0.07 / 2 = 0.035 per side. The length of such a cut should be sufficient for micrometer (0.35 in the example). Also, the machining speed and cutting feed rate should be the same for best results.

The trial cut is a special cut performed on demand only. Although a lathe example has been presented here, trial cuts have many applications in milling operations as well. The main key is to identify the need first; then, provide a suitable program.

Easing Up on Calculations

April 2008, updated February 2013

In life, there are some inevitable facts we have to live with. In manual CNC programming, there are many as well, but one stands above them all — one that is often dreaded by new programmers and cursed by those more experienced — and that is math. A surprising number of students who attend various CNC courses are not aware that math (or to be more precise, mathematical calculations) is an integral part of being a CNC programmer who does not have an access to CAM-based software. This fear of math, or at least this apprehension of math, stands on a rather shaky ground. True, manual CNC programming does require calculations. Math is, of course, at the center of such calculations. As a scientific subject, the field of mathematics is colossal. Yet, in virtually every field, there is no need to know all math offers. In fact, math applications are always limited to a particular field, and it is up to the user to learn them and use them correctly.

A Reader for Programmers

Arithmetic and Algebra

CNC programming is no different. The inherent trepidation that somehow affects new programmers can be easily dissolved by better understanding of what part of the huge subject of mathematics is actually needed for every day work. To categorize math subjects as they relate to CNC programming is not difficult, as there not that many of them. First and foremost are actually two subjects, equally important: arithmetic and algebra. The word arithmetic sounds special, but there is nothing special about its meaning. Arithmetic is simply the part of mathematics that deals with the four basic functions of addition, subtraction, multiplication, and division. We all use these daily, without even thinking twice about them.

Algebra is positioned a step higher, but it's also very common in everyday life. Algebra involves working with some known values to find unknown values. This work is often expressed in formulas, where an unknown value can be calculated from some known values. A typical example of a simple algebraic function is the common conversion between inches and millimeters. If one inch = 25.4 mm, then two inches will be 2 x 25.4 = 50.8 mm. The formula is mm = inches x 25.4. This formula demonstrates algebra at its basic level.

Basic Formulas, Geometry, and Trigonometry

In CNC programming, you frequently work with only one or two dozen basic formulas. They relate to calculations of spindle speeds, feed rates, cutting time, threading data, etc. They also relate to some geometry calculations, for example, calculations of tapers. Special situations require special formulas, but not necessarily more difficult formulas.

Not all geometry calculations can be made with algebraic functions, and that is where the third level of math comes in: geometric calculations. This area of math is huge by itself, but in CNC programming, only a very small part is required. This part includes the basic trigonometry needed to solve right angle triangles. Add to it related subjects such as the Pythagorean Theorem and the Law of Similar Triangles, and you have just covered virtually all calculations needed for programming standard mills, machining centers, and lathes.

Not that long ago, a scientific pocket calculator was about the only aid available to CNC programmers. Although the need for a calculator has not changed in many cases, it is far from being the only aid. In fact, there are many other resources that assist with solving math related problems, even without a pocket calculator. A computer-based calculator is simple, but is accessible to anybody with a personal computer. Most major cutting tool companies provide on-line or downloadable software for calculating various cutting conditions, even complete solutions, for example, a thread milling software — and that software is usually free of charge. Of course, such software is geared towards a particular tooling brand, but in many cases it can be used for any brand.

On a commercial level, if you have spreadsheet software such as Microsoft Excel, you can use it in many areas of CNC programming, including automation of tasks such as bolt hole circle calculations. Yes, it requires some development effort, but the time may be well spent. Several companies offer CNC specific software in the form of practical utilities, from editors to graphic back plotters. One software program that offers a ton of CNC utilities is the excellent, inexpensive, and very popular Machinist Toolbox by Tim Markoski. It not only removes any need for a pocket calculator but also provides a friendly graphical interface for hundreds of calculations and instant solutions in the area of CNC programming and machining.

Not every small shop can afford CAM software and not every small shop may even need it. Manual programming may have its many disadvantages, but with computer-based software and many available utilities, the venerable pocket calculator does not have to do all the work it once did so gracefully.

A Reader for Programmers

 ## Preventing Scrap with Offsets
May 2008, updated February 2013

The major goal of a CNC operator should always be to make any part to drawing specifications. At the same time, the operator should also make an effort to prevent making a scrap. It is not uncommon to scrap a few small pieces made of a long bar before making the perfect part. This leisurely approach often leads to complacency and may prove costly when the number of blank parts is fixed.

Methods for Preventing Scrap

There are only a few reasons for scrap to happen in the first place. On the human side, it is either negligence or incompetence. On the machine side, it is a hardware or software failure. Needless to say, the human errors happen more often. There are several ways CNC programmers and operators can prevent a scrap. Some procedures have to be included in the program; others can be done at the machine directly. Programmers often include extra features in their programs that help to eliminate scrap. Typical methods are:

- Providing a trial cut, using block skip function or some other method
- Including *Program Stop/Optional Stop* functions M00/M01
- Including messages or comments
- Making a setup sheet with a sketch
- Splitting a tool motion for single block operation
- Programming sufficient clearances

These are just some of the methods that can be built into the program to make it easier for the operators. CNC operators also have several ways to prevent a scrap created from any cause. They must identify a potential problem before it becomes a real problem. The first goal is to be able to interpret the program and know what it does. Mistakes made during tool or fixture setup are a common cause of scrap. Errors also happen during data

CNC Tips and Techniques

entry into the control system. Placing the right settings into a wrong offset register is a quite common cause. Also common are errors related to decimal point entries, either as a missing decimal point or a decimal point in the wrong place. Even after a few parts have been machined successfully, scrap can happen when changing offsets or tools or inserts, and so on.

Measured Part Dimensions

Even a verified program is no reason to decrease the operator's attention. When working with offsets, there are ways to let them work for you. The following evaluations show how a measured part dimension influences the whole part and can be used for external or internal contours:

For both external and internal contours, if the measured dimension is within tolerance, no action is necessary. An undersize external contour is a scrap; an oversize contour can be re-machined. An undersize internal contour can be re-machined; an oversize dimension is scrap.

Considering these facts, the operator can take precautions by intentionally manipulating cutter radius offset. The following program does nothing more than makes an external contour to machine a small rectangle 75 x 50 mm with a 10 mm end mill:

```
N1 G21
N2 T01 M06 (10 MM END MILL)
N3 G90 G54 G00 X-7.0 Y-7.0 S1000 M03
N4 G43 Z2.0 H01 M08
N5 G01 Z-5.0 F500.0
N6 G41 X0 D51
N7 Y50.0 F150.0
N8 X75.0
N9 Y0
N10 X-7.0
N11 G40 G00 Y-7.0 M09
N12 G91 G28 Z0 M05
```

A Reader for Programmers

N13 M30
%

Assume tolerance on the 75 mm width to be tight, zero to 30 microns plus. Even with a brand new cutter, the width will be right on or slightly under. The width may be larger, too. Rather than taking a chance, use the following procedure:

1. Identify the normal offset value for D51.
2. Estimate the amount of possible error.
3. Temporarily increase the offset.
4. Cut, then measure and adjust offset.

Here are the details. For *Item 1*, the answer is 5.000 mm for a 10 mm end mill. *Item 2* needs special attention so the estimate is good. Tool quality is important — a new cutter should be within 10–15 microns. Consider this for *Item 3:* changing the offset, so the part will be wider than 15 microns. If stock size allows, you may go with 200 microns on the width. What will the temporary offset setting be?

From the original evaluation, a larger external width can be re-cut to final size. It means adding stock to the tool path by increasing offset from 5.000, but to what? Achieving the goal of width that is 200 microns larger means only 100 microns per side! The offset has to be set to 5.100, not 5.200! Upon completion of Item 3, D51 setting will contain a temporary 5.100 mm offset.

Next is *Item 4*. When the cutting is completed, measure the 75 mm width. It can be exactly 75.2 mm, or it can be larger or smaller. Even if the width is undersize, it still should be greater than the 75.03 maximum. If it is not, you made a mistake in *Item 2* — estimating the error amount. Let's look at the outcomes and offset adjustments, but first you have to decide what the final width should be, for example, the middle 75.015 width. Measured sizes are only examples found on the next page:

CNC Tips and Techniques

ON SIZE result = 75.200 = (75.015 − 75.200) / 2 = −0.0925.

Use 0.093 to match three decimal places. Subtract 0.093 from the current setting of the offset 51: 5.100 − 0.093 = 5.007 as the final D51 setting.

OVERSIZE result = 75.243 = (75.015 − 75.243) / 2 = −0.114, and 5.100 − 0.114 = 4.986 as the final D51 setting.

UNDERSIZE result = 75.185 = (75.015 − 75.185) / 2 = −0.085, and 5.100 − 0.085 = 5.015 as the final D51 setting.

One lesson can be learned. Establishing the final offset amount is the same, regardless of the actual size of the test measurement. The ultimate goal should be to eliminate scrap, not to minimize it.

Interpreting a CNC Program
June 2008, updated February 2013

The ability of CNC operators to interpret a part program is one of their most important skills. Unfortunately, it also happens that not every program reaching the floor is flawless. Programming errors do happen, and it is often up to the operator to find them and even debug them. Being able to interpret the program *(even without a drawing)* is the first step of being able to fix the errors, if necessary. Take a look at this example, which has no errors:

O1001 (ELEVATION PLATE)
N1 G21
N2 G17 G40 G80 T07 (T7 = 12.7 MM DIA SPOT DRILL)
N3 M06
N4 G90 G54 G00 X20.0 Y60.0 S900 M03 T10

A Reader for Programmers

```
N5 G43 Z25.0 H07 M08
N6 G99 G82 R2.0 Z-2.6 P200 F200.0 (0.35x45 CHAMFER)
N7 X30.0 Y100.0
N8 X100.0
N9 Y60.0
N10 G80 Z25.0 M09
N11 G28 Z25.0 M05
N12 M01
```

When evaluating a part program, some information is included in the comments, for example, the part name. Less obvious is the size of the tool used for this operation — here, it is a half-inch spot drill, converted to millimeters. This method is a common one in many machine shops, where a metric part uses imperial tools. What else can be gathered from the program?

Units

For any job, it is always important to know the dimensional units. In a shop that works with both metric and imperial units, there are two methods to find out which one is active. One is by the program codes G20 (inch) or G21 (metric). Another method is to find out at the control, usually under Settings. With G21 in block N1, this example is in metric units.

Type of Operation

CNC operator should be able to find out what *type of operation* is taking place from the program itself. Prevailing use of G01, G02, and G03 commands suggests a contour or a pocket machining. Using XYZ axes in G01 mode for most of the tool suggests 3D machining. For holes, various fixed cycles are used, such as the G82 cycle in the example for spot drilling. Also associated with fixed cycles are retract commands G98 and G99. Spot drilling can be further identified by a rather shallow cutting depth, dwell (P), and the existence of a *drill* as the next tool in the program.

CNC Tips and Techniques

Speeds and Feeds

For a safe machining operation, identifying the spindle speeds and cutting feedrates in the program helps to make intelligent decisions when using spindle and feedrate overrides, particularly for the initial part. As an operator, keep in mind that even the most experienced programmer may not always be exact with speeds and feeds. In fact, optimizing speeds and feeds at the machine is often one of the most important functions of a CNC operator.

This example uses one spindle speed of 900 rpm (S900) and one feedrate of 200 mm/min (about 8 ipm). Some milling jobs may have two or more feedrates applied to the same tool.

Spindle rotation is also important. When interpreting a part program, three M-functions relate to spindle rotation: M03 (CW), M04 (CCW), and M05 (spindle stop). M03 is the standard rotation for right hand tools whereas M04 is used for left hand tools.

Tool Numbers

Many errors at the machine happen because of wrong tool selection. In the example, the tool number is identified by the address T. Its meaning is determined by the type of automatic tool changer (ATC). For the *fixed type ATC*, each tool returns to the same magazine pocket it came from; the address *T* indicates both the pocket number and the tool number. In case of the more common *random type ATC*, the address T identifies the tool to be used next.

There are two T addresses in the ex.: one in block N2 (T07) and the other in block N4 (T10). Note that the block N3 contains ATC function M06. Once the tool change has been completed, the control searches for T10 to be ready in the magazine. To interpret these three blocks, T07 *means get tool number 7 ready for tool change,* M06 means *get T07 to the spindle,* and T10 means *get the next tool T10 ready.* Having the next tool ready is one of the greatest advantages of the random type automatic tool changer.

A Reader for Programmers

Dwell Time

Once a dwell is identified in the program — *P250 in block N6* — make sure you understand the reason for the dwell. Generally, pay special attention to dwells longer than one half of a second. Longer dwells are not necessarily a waste of time, if used properly. A long dwell is often programmed to make sure a certain mechanical function of the CNC machine is fully completed before program processing continues. For example, activities such as clamping and unclamping, automatic door closing, delay due to acceleration, tailstock application, and others, benefit from longer dwells. In these cases, you should *never* change the dwell time. For machining, the dwell time change (usually a reduction) is quite often justified.

Start with minimum dwell calculation (60/rpm). Minimum dwell is the time required to complete one spindle revolution. In the example, 60/900 = 0.067 seconds, 67 ms (milliseconds), is the minimum dwell. Programmers often double or triple the minimum dwell to allow for the possibility that spindle speed override is set to 50% (minimum override). P200 in block N6 allows fewer than three revolutions while the dwell is in effect.

Other discoveries can be made by careful interpretation of the program.

Now, a challenge question: Can you tell the *diameter* of the drill that follows the spot drill? No other tools will be used. See the answer and explanation in Appendix 1 at the end of this book.

CNC Tips and Techniques

 ## Default Settings in Macros
July 2008, updated February 2013

In spite of the great advancement of CAM software, user macros (custom macros) continue their special place in CNC programming and their use is on the rise. They are used in addition to — *not as a replacement of* — manual, conversational, and computer assisted CNC programming. Macros are used to provide special solutions to special requirements. In simple terms, a macro is a type of subprogram with many added features, particularly its ability to manipulate variable data. Several control manufacturers, such as Fanuc, Okuma, Fadal, and others, offer macros either as a standard or an optional feature.

Overall, Fanuc macros, called *Custom Macro B*, are the most widely used. The typical application in a program is to call a macro by its program number with preparatory command G65, followed by one or more arguments. An *argument* is a specific definition that is passed to the macro body. Different data in the arguments force the macro to behave differently, while fully under the programmer's control.

Bolt Hole Circle Macro

One of the classic examples of a macro is a *bolt hole* circle macro. There are many variations of this macro, but generally they share a minimum of two variable data: *bolt circle diameter* and the *number of holes*. This is the most basic requirement, but it does not make the macro very flexible. For example, the bolt circle has to be located at a certain XY coordinate position; the angular orientation of the bolt pattern may be specified in the drawing. If the additional data are not specified, the macro can only be used at one fixed XY position, for example X0Y0. Also, such a macro will have angular orientation fixed as well, usually at zero degrees. For many bolt holes, the two basic definitions may be sufficient. However, a well-developed macro should

A Reader for Programmers

have flexible XY location as well as a flexible angular orientation built-in.

As an example, take the following bolt pattern macro call (can be used for both metric and imperial dimensions):

G65 P8500 X0 Y0 D49.0 H6 A0 S1

The X and Y arguments (#24 and #25) are locations of the bolt circle center. Argument D (#7) is the bolt hole circle diameter, H (#11) is the number of equally spaced holes, A (#1) is the orientation angle measured from zero degrees. Adding the S (#19) argument, starting hole number can be changed and one or more initial holes skipped. In the parentheses are fixed variable numbers associated with each argument — they are part of the macro body. The G65 example shows a typical entry for most bolt holes. The slight disadvantage is that once the XY, A and S arguments become part of the macro, they have to be defined in the G65 statement. They have to be repeated from one macro call to another, even if they are always the same. *Unless...*

... unless they are defined as defaults. Default in a macro program is a condition that will assume a certain value or amount without actually specifying it. In our case, we want to set X and Y arguments to zero (X0Y0), angular orientation at zero degrees (A0), and starting at the first hole of the pattern (S1). This can be achieved by a simple conditional test, using the **IF** function checked against existence of an argument. For example, if the X argument is missing, assume it is zero, and so on. Fanuc macros offer a special *null variable* identified as **#0**; this variable has no value, it is empty. A series of statements can be placed in the macro body that will check for existence of a particular variable, and if that variable if not defined in the argument list, macro will set it to the desired value. In order to allow other settings, a **GOTO** statement is used to bypass the default definitions. Here is a set of the four defaults:

CNC Tips and Techniques

```
IF[#24 NE #0] GOTO10
#24 = 0 (if X is not defined, set X to X0)
N10 IF[#25 NE #0] GOTO20
#25 = 0 (if Y is not defined, set Y to Y0)
N20 IF[#1 NE #0] GOTO30
#1 = 0 (if A is not defined, set A to A0)
N30 IF[#19 NE #0] GOTO40
#19 = 1 (if S is not defined, set S to S1)
N40 ... (macro continues)
```

If your control system supports the IF-THEN statement, the above blocks can be replaced with a shortened conditional testing:

```
IF[#24 NE #0] THEN #24 = 0 (if X is not defined, set X to X0)
IF[#25 NE #0] THEN #25 = 0 (if Y is not defined, set Y to Y0)
IF[#1 NE #0] THEN #1 = 0 (if A is not defined, set A to A0)
IF[#19 NE #0] THEN #19 = 1 (if S is not defined, set S to S1)
```

In the following example, the XY center location is defined at a location other than zero and will override the macro default. On the other hand, A and S arguments are missing (null), and their respective defaults will be used:

G65 P8500 X50.0 Y37.5 D49.0 H6

For many bolt circle patterns, where the XY is a zero location as well, the macro call can be simplified further to:

G65 P8500 D49.0 H6

This method shows the power of using defaults in custom macros. See Appendix 2 at the back of this book for a complete working macro listing and accompanying figure (Figure A2-1).

A Reader for Programmers

 ## Create Your Own G-Code
August 2008, updated February 2013

Custom macros are a very common method of enhancing the concept of subprograms by adding variable data and many data manipulation features. One of the less frequently used macro features involves the development of customized G-codes. Typically, a custom G-code for machining will be some kind of a cycle and will appear as such in the program. G65 statement can also be used, but it does not provide the same transparency. The macro example used in this essay will recreate the G13 command used on some controls other than Fanuc. As the actual control model is very important, the macro will be developed for Fanuc 16/18/21 systems.

The G13 Command

Preparatory command G13 is commonly used to machine a circular pocket, such as a counter bore or a spot face, in climb milling mode. The macro will be stored as a cycle using only three major tool motions:

- Motion 1: Lead-in arc —180 degrees from the circular pocket center
- Motion 2: Full circle machining
- Motion 3: Lead-out arc —180 degs. back to the circular pocket center

As cutter radius commands G40/G41/G42 cannot be applied on an arc, they will not be used. Yet, the macro will still allow the CNC operator to fine tune the pocket dimensions at the machine, following established procedures as with a cutter radius offset. Development will start the same way as for any G65 macro call. Three variable arguments will be required:

- Circular pocket diameter — argument C (#3 in macro)
- Tool offset number to store the cutter radius — argument D (#7 in macro)
- Cutting feedrate — argument F (#9 in macro)

CNC Tips and Techniques

In the first step, develop a standard macro, for example O8600:

O8600 (G13 CIRCULAR MILLING CYCLE — CLIMB MILLING)
#31 = ABS[#3]/2 *Radius of the circular pocket — guaranteed positive*
#11 = #4001 *Stores current G-code of Group 01 (motion commands)*
#13 = #4003 *Stores current G-code of Group 03 (absolute/incremental)*
#32 = #31–#[2000+#7] *True radius of the pocket to cut; see explanation below !!!*
IF [#32 LE 0] GOTO998 *Error if radius offset value is too large*
#33 = #32/2 *Calculated lead-in/lead-out radius*
G91 G03 X#32 I#33 J0 F[#9/2] *Motion 1 — lead-in arc toolpath*
I-#32 F#9 **Motion 2** *— full circular pocket toolpath*
X-#32 I-#33 J0 F[#9/2] *Motion 3 — lead-out arc toolpath*
G#11 G#13 *Restore original G-codes of groups 1 and 3*
GOTO999 *Bypass error message*
N998 #3000 = 13 (OFFSET TOO LARGE) *Alarm refers to actual offset setting*
N999 M99 *Macro end*
%

Once available, the macro can be called along with all three arguments:

G65 P8600 C60.0 D51 F200.0 *(metric units)*

Be sure to enter the D argument as an offset number, not the cutter size!

Converting Macro Call G65

The next step is to convert the macro call G65 to a new G-code (G13). This choice is suitable because this type of macro is a true machining cycle. To make it appear as a cycle in the program, a G-code that has not been assigned must be selected. To make the macro a true cycle, two changes are required. The first change must be selection of a special macro number from the allowed range of ten possibilities: O9010 to O9019. That means

A Reader for Programmers

changing the macro number from O8600 to O9013, for example. Any number within the range can be selected, but calling the macro O9013 makes some logical sense. The second step is to register the new G-code (13) as a system parameter.

Offset number 13 has to be registered in the parameter that corresponds to the calling program, which is O9013. Depending on the control model used, the range of system parameters will be different (consult your control manual). This example uses Fanuc 16/18/21, where the parameter range is between 6050 and 6059 (relative to macros 9010 to 9019). Parameter 6053 corresponds to macro O9013 and will contain the new G-code number 13. Once the new G-code number is registered, the macro body does not change, only its number. Now, a circular pocket can be called by the new G-code in any part program that requires it:

G13 C60.0 D51 F200.0

Types of Macros

The example used Fanuc system model 16/18/21, but it also used offset memory *Type A*. This is reflected in variable #32. Macro variable #32 is absolutely critical and requires explanation. All active offsets are stored as system parameters. Fanuc offers three types of memory offset registers (*Type A* — one register is shared by tool length and radius offsets; *Type B* — also a shared register, but geometry and wear offsets are separate; *Type C* — separate registers for tool length and radius offsets, each with geometry and wear settings). For 200 offsets or less, *Type A* macro uses system variables #2001 to #2200, *Type B* uses system variables #2001 to #2200 for the geometry offset, and #2201 to #2400 for the wear offset. Finally, *Type C* uses system variables #2401 to #2600 for the radius geometry offset, and system variables #2601 to #2800 for the radius wear offset. Variable #32 must always reflect the offset memory type used. There are no tool length system variables for types *A* and *B*, and those for *Type C* are not required. Various error trapping features can be added

CNC Tips and Techniques

to the macro, which can also be modified to conventional type of machining.

The macro listing for the example with an accompanying figure appears in Appendix 3 at the end of this book.

Scaling Option
September 2008, updated February 13

One function on many control systems is called scaling. Often offered as an optional control feature, scaling allows an existing program to machine a part smaller or larger. Before you jump to conclusions, keep in mind that scaling is not meant to be the means of part correction due to poor programming methods or bad setup. Scaling is often used in casting or plastic injection mold work to allow adjustments due to shrinkage during the molding process, but can be used in 2D work as well. From a CNC programmer's perspective (rather than engineering perspective), scaling is quite easy to program, but there are a number of some special considerations.

Changing the Toolpath

Scaling will physically change the current toolpath proportionally, based on the location of the scaling center and the scaling factor provided. A control-dependent system parameter can be set to make scaling effective or ineffective for each of the three main axes, but not for any additional axis. Also, scaling does not affect various offset modes, namely work offsets G54–G59, tool length offset G43, and cutter radius offset mode G41–G42. If a fixed cycle in the G81–G89 series; or G73, G74, or G76; is used while scaling mode is active, the shift amount in G76 and G87 cycles (the peck drill depth and relief amount in G73 and G83 cycles) will *not* be affected.

A Reader for Programmers

The *machine zero return* commands are another consideration. Preparatory commands related to the machine zero return, particularly G28 and G30, but also G27 and G29 (rarely used anymore), should always be programmed with scaling mode OFF in effect. On older controls that still use the G92 command for position register, make sure the scaling mode OFF is also in effect. Cutter radius offsets G41 and G42 should always be canceled by G40 before the scaling function is activated. As mentioned already, other commands and functions can be active, including all work offset commands G54 through G59 and the additional ones, if they are available.

G50 and G51 Codes

Fanuc programming format uses two G-codes: G50 and G51 commands, where G50 cancels the scaling mode and G51 makes the scaling mode active. The programming format for G51 is:

G51 I.. J.. K.. P..

IJK addresses refer to their respective XYZ coordinates of the scaling center, while the P-address is the scaling factor. Scaling center is defined in absolute coordinates of the current units of measurement, and scaling factor can be in 0.001 or 0.00001 increments, depending on the control setting. In order to prevent any conflict, both G50 and G51 should be programmed as separate blocks, with no other significant data included.

The scaling center determines the *location* of the scaled tool path and the scaling factor determines its size. For scaling factors greater than one, the machining tool path will be *expanded*. For scaling factors smaller than one, the machining tool path will be *reduced*. The scaled part will always expand away from or reduce towards the scaling point equally along all specified axes.

The basic programming format is illustrated in the following example:

CNC Tips and Techniques

N1 G20
N2 G17 G40 G80
N3 G50 (SCALING OFF AT TOOL START)
N4 G90 G00 G54 X-1.25 Y-1.25 S800 M03
N5 G43 Z1.0 H01 M08
N6 G51 I0 J0 P1.050 (SCALING CENTER AT X0Y0 - 1.05X expand)
N7 G01 Z-0.7 F50.0
N8 G41 X-0.75 D51 F25.0
N9 Y1.75 F15.0
N10 X1.5
N11 G02 X2.5 Y0.75 I0 J-1.0
N12 G01 Y-0.75
N13 X-1.25
N14 G40 Y-1.25 M09
N15 G50 (SCALING OFF AT TOOL END)
N16 G00 Z1.0
N17 G28 Z1.0
N18 M30
%

Rounding Errors

One very important consideration in scaling is a possible rounding error. To illustrate, consider a conversion between metric and inch units. A dimension of 1.0 inch will become 25.4 mm, which is an exact conversion. On the other hand, 25.4 mm converted to inches will use a factor of 0.039370079. If the scaling factor is set to a three decimal accuracy, the 25.4 mm will become only 0.991 inches (25.4 x 0.039 = 0.9906) — a rather large error of about 0.009. With scaling factor set to five decimal places, the error will still be there but it will be practically negligible.

The scaling factor is a *multiplying factor*, not a percentage. If a certain scaling percentage is required — for example, 5% shrinkage — it has to be converted. The scaling factors of P1.05 (expansion) or P0.95 (reduction) are well within the expected accuracy of the final part precision.

For a complete example with illustration, using the scaling function, see Appendix 4 at the back of this book.

A Reader for Programmers

Safety and CNC Programming
October 2008, updated February 2013

When it comes to safety in machine shops, the great majority adheres to the occupational safety laws and regulations of the jurisdiction in which they are located, often exceeding the official ones. Common rules are generally well known. They include wearing protective clothing, safety boots, safety glasses with protective side shield, keeping the environment clean, removing any hazards, and many others. Machining safety is paramount to the successful operation of any machinery, CNC equipment included. Many safety issues are associated with the actual equipment operation and machining of a part. Very few companies have standards relating to safety issues in CNC programming, whether manual or computer generated. CNC programmers may work from an office environment, yet they have a great responsibility to build safety features into every CNC program.

A fairly long list can be assembled to identify what such features should be. A lot depends on the product manufactured, the equipment used, and company standards and culture. Certain general approach to safety in CNC programming can be identified, regardless of local conditions or method of programming. A short list focused on CNC vertical machining centers can be adapted to CNC lathes and other machines, and may include the following items:

Approach program development with safety in mind for all work.

This underlying philosophy should be constant for any part, any program, and any machine.

Program your approach towards a part in the XY axes first; then move the Z-axis.

When the tool moves above the part first, it is less likely to come into collision with the part itself or the fixture. In terms of

CNC Tips and Techniques

efficiency, it takes longer than three-axis simultaneous motion, but this XYZ split motion can always be changed after a safe setup had been established.

Program a retraction from a part in the Z axis first, then move the XY axes.
Tool retraction from a part may include only the Z-axis, but for XYZ motion, the split is safer than three-axis motion. This is the opposite of the previous item.

Start every program with an initial setting and cancellations (so-called status block or a safety block).
It is usual to include various settings and cancellations at the beginning of a program. Settings may include dimensional units selection, plane selection, and absolute mode whereas cancellations typically include fixed cycles, cutter radius and length offset, and initial status of certain options (for example, scaling off or polar coordinates off).

Provide sufficient clearances (above part, below part, lead-ins, and lead-outs, etc.).
Insufficient clearances are often the cause of dangerous tool-machine collisions. Rapid motion to a certain depth may be safer if programmed as two motions: above the part first, then to the required depth. This method helps during setup and does not increase cycle time during automatic operation.

Program reasonable speeds, feeds, depth of cuts, width of cuts, etc.
Every CNC machine has override switches — establishing speeds and feeds on the conservative side may my safer, although less efficient. Efficiency can be achieved by optimizing program at the machine during setup, using the overrides. This approach may be particularly useful when working with less experienced operators.

A Reader for Programmers

Select a safe tool change or part change position.

To make automatic tool changes on vertical CNC machining centers, the tool has to be positioned at the Z-axis machine zero. The common approach is to make the tool change above the current tool location. For long tools or high setups, this location may not be safe and should be made off the part.

Select suitable tools for the job.

Incorrect tools for the job are not only a common cause of delays, but they are also the cause of unsafe working conditions. Describe the tool correctly, for example, as a *center-cutting end mill* rather than just an *end mill*.

Program tool motions towards a solid support, such as the fixed jaw of a vise.

Physical forces are at work during any machining. Programming the direction of a cut to prevents the part from even a slight motion is always important.

If a program fails, correct the REAL cause of the problem, not a perceived one.

Don't play the blame game — nobody wins it. If the error is in the program, it is the programmer's fault; it is as simple as that. It is not the job of CNC operators to find and correct program errors.

There are many more items that can be added to this short list. For more information, see Appendix 5 at the back of this book.

CNC Tips and Techniques

 ## Special Tapping Macro
November 2008, updated February 2013

A while back I trained a customer in programming and operation of his new CNC machining center with a Fanuc control. His product was quite simple, but included a lot of tapped holes in aluminum. These holes were very small (the largest was #3-48), cutting through material no thicker than 1/8 of an inch (often much less). I might add that the customer was very experienced in manual machining. Yet, what looked like a short training session turned out to be a two-day job. The problem was not uncommon to tapping with very fine pitch: *thread stripping*.

Thread Stripping

We tried everything that our collective experience has ever taught us and more. We had the machine checked by the factory technician, used different type of floating tap holders (no rigid tapping was available), and tried taps with different geometry, flutes, and coating. We tried different spindle speeds, coolants, and tap drill sizes — all to no avail. Giving up was not an option, so I consulted a tooling expert. To all questions *"Did you try this?"* I answered with "Yes," except for one question. *"Did you try to feed in at 80% and out at 120%?"* I answered I did not because the G84 tapping cycle does not allow change of feed rate. We did try underfeeding, but not different feed rates. After a short lecture on the benefits of this rather unusual method, I wrote a test program and tried it at 100% spindle speed and feedrate setting. To everyone's surprise, it worked. Changing the test program to a macro was the next step:

(#3-48 TAP)
...
S800 M03
G65 P8500 R0.15 Z-0.14 F16.6 (MACRO APPLIED WITH ARGUMENTS)
...

A Reader for Programmers

O8500 (SPECIAL TAPPING MACRO)
G90 G00 Z#18
G01 Z-[ABS[#26]] F[#9*0.8] M05   (FEED IN AT 80 PERCENT)
Z#18 F[#9*1.2] M04   (FEED OUT AT 120 PERCENT)
M05
M03
M99
%

System Variables

This macro represents only the main core. As is, it lacks certain important features that are inherent to a standard G84 cycle:

- G84 always works at 100% feed rate, regardless of the override switch setting.
- Single block mode is disabled.
- Feedhold is disabled.

Fanuc provides two system variables for this purpose: #3003 and #3004. Variable #3003 controls single block mode — *0=Enabled, 1=Disabled* — and it waits for completion of MTS function before executing the next block (there are also settings 3 and 4). Variable #3004 controls feedhold, feed rate override, and exact stop check with eight variations available — *0=All Enabled*. To disable both the feedhold and feed rate override, but not the exact stop check, the setting must be #3004=3. When the macro exits, a normal setting should be established:

O8500 (SPECIAL TAPPING MACRO)
IF [#18 EQ #0] GOTO 991
#3003 = 1   (SINGLE BLOCK MODE DISABLED)
G90 G00 Z#18
#3004 = 3   (FEEDHOLD AND FEEDRATE OVERRIDE DISABLED)
G01 Z-[ABS[#26]] F[#9*0.8] M05   (FEED IN AT 80 PERCENT)
Z#18 F[#9*1.2] M04   (FEED OUT AT 120 PERCENT)
#3004 = 0   (FEEDHOLD AND FEEDRATE OVERRIDE ENABLED)
M05

CNC Tips and Techniques

```
M03
#3003 = 0  (SINGLE BLOCK MODE ENABLED)
GOTO 999  (BYPASS ERROR MESSAGES)
N991 #3000 = 991 (FEEDPLANE R NOT DEFINED)
N992 #3000 = 992 (Z-DEPTH Z NOT DEFINED)
N993 #3000 = 993 (FEED RATE F NOT DEFINED)
N999 M99
%
```

The macro shows not only the use of several 3000-series system variables, but also how a relatively simple but pesky little problem can be solved. Further improvements to the macro can be made; for example, the feed rate can be calculated internally, based on the spindle speed and threads per inch or pitch.

There is also another point to this story — take advantage of the knowledge many tooling specialists have.

Setting Up a New Part
December 2008, updated February 2013

CNC machine operators usually start a new job by studying the documentation included with the program, using the engineering drawing as the main source of information. Additional data are often covered in a setup sheet or a tooling list, along with a graphic representation of the setup. This month we look at general setup procedures that can be used for various types of CNC machines.

Step 1: Set the cutting tools.

This first step uses tooling information from the part program. CNC operators set the cutting tools into their holders and respective tool stations and register all tool numbers into the control memory, if necessary. Make sure the tools are sharp and mounted properly in the holders.

A Reader for Programmers

Step 2: Set the fixture.
A fixture that holds or supports the part is mounted on the machine, squared, and adjusted, if necessary, but the part itself is not mounted at this point. The setup sheet serves as the documentation, particularly for complex setups. A fixture drawing may often be required as well.

Step 3: Set the part.
Locate the part into the fixture and make sure it is secured. Check for possible interferences and obstacles in the setup. Obstacles generally include undesired contact of the tool with the part, machine, or fixture. Watch for clearances during the tool change.

Step 4: Set the tool offsets.
Depending on the type of machine, this step takes care of setting the tool geometry and wear offsets, tool length offset, and cutter radius offset, if applicable. One of the most important parts of this step is the setting of work coordinate system (work offsets G54 to G59) or the tool position registers (G92 or G50) for old controls, but not both, even if they are available. Work offset setting *(geometry offset for lathes)* is by far the best and most convenient selection of modern CNC machine tool setup.

Step 5: Check the program.
This step is the first evaluation of the part program itself. The part to be machined may be temporarily removed from the fixture. Because all the offsets are already set in the control system, the program is checked accurately, with all usual considerations. Program override switches on the operation panel may be used, if required. Watch for tool motions in general and be sure to watch for tool indexes specifically. Repeat this step, if not absolutely sure about any aspect of the programmed toolpath.

CNC Tips and Techniques

Step 6: Reset the part.
If the part was removed in the previous step, now is the time to mount it in the fixture again. Successful completion of all previous steps allows you to continue proving the first part. At this point, check the tooling once more; also check the oil and air pressure, clamps, offsets, switch settings, chucks, and any other important machine feature, just to be sure.

Step 7: Make a trial cut.
Sometimes an actual trial cut may be required in order to establish whether or not the programmed speeds and feeds are reasonable and whether the various offsets are set properly. The trial cut is a temporary or an occasional cut that is designed to identify minor deviations in actual offset settings and allows their change. Make sure the trial cut leaves enough material for actual machining. Trial cuts also help to establish tool offsets to keep dimensional tolerances within limits.

Step 8: Adjust the setup.
At this point, any necessary adjustments are finalized in order to fine tune the program before actual production begins. This step includes final offset adjustment (usually a wear offset). It is also a good time to adjust spindle speeds and cut feedrates, if necessary.

Step 9: Start the production batch.
A full batch production can start now. A quick second double check may again prove to be worth the time. The ideal way to run a new program is to run it first through the control graphic display, if available. It is fast and accurate, and offers a lot of confidence before actual machining. This test can be run with a variety of override modes in effect, for example, *Machine Lock or Single Block*. Do not underestimate features such as *Zero Axis Neglect and Dry Run*, when testing a new program. Do not expect

A Reader for Programmers

100% accurate graphical display of any toolpath. No display can show every single detail and no simulation will show flying chips.

Although the numbering of individual steps described suggests a certain fixed order, there is a great amount of flexibility in the process. Just use common sense — past experience also helps.

Readings From 2009

A Reader for Programmers

 ## Getting Rid of Chatter
January 2009, updated February 2013

Every machinist has experienced the sound — sharp, squeaky, a bit eerie, but definitely spelling trouble. This is the sound of chatter that leaves nothing to imagination. Every machinist (CNC or conventional) will know that something is not right with the cutting tool performance. Even if the sound itself were tolerable, its results are not. Chatter leaves a poor surface finish on the part, mostly unacceptable, and negatively influences dimensional tolerances. Chatter also shortens the tool life, often quite dramatically. On a positive note, chatter can be generally controlled to the point of total elimination. Chatter during machining is a type of undesired vibration, caused by the cutting tool or the part working against each other, rather than with each other. The actual cause of such vibration is not always easy to identify quickly, at least not without some special equipment. It is not unusual to find that the actual remedy may take a few trials to succeed. However, there are quite a few options the CNC programmers or the machine operators can explore and apply to eliminate chatter in most cases.

Factors Behind Chatter

Chatter is more often present on CNC lathes than on machining centers, particularly during deep-hole boring operations. A boring bar can work very well if the ratio of its overhang length to its diameter is within reasonable limits. Normally, the ratio is about 3:1 for steel shank boring bars, and 5:1 (or more) for carbide bars. In practice, it means that a 25-mm steel shank bar (1.0 inch) can be extended about 75 mm (3 inches), or 125 mm (5 inches) for a carbide bar.

Actual cutting conditions, such as the material type, depth of cut, cutting speeds and particularly feedrates also play a large role. Long bar overhang is one of the most common reasons for chatter to occur. Although the solution is simple — just shorten the bar — this is not always a feasible solution, due to the part design and the necessity of a long boring bar. Choosing a carbide bar, rather than a steel bar, often helps.

Another alternative is to purchase a special boring bar designed to dampen the vibration. Some homemade dampening bars can be made from steel bars with a drilled cavity filled with large size bearing balls (or carbide balls). This can also be an inexpensive option for some jobs. Otherwise, a change of setup, increasing the number of operations and/or tools may be the last resort. If the boring bar overhang is not the culprit, the problem may be overhang of the part itself.

The part extended from the chuck also has its own limitations. Fortunately, an extended part can be supported by means of tailstock and/or steady rest or follow rest. Boring operations are not possible with supporting tailstock, and the alternative of a steady rest or follow rest attachment can provide the solution. Always make sure the boring bar is properly set in relationship of the tool tip and the part center line.

Other common causes of chatter can be traced to a weak setup. Small depth of grip, insufficient support in the holding device, and wrong chip breakers should also be looked into as possible causes. In some less common cases, the machine tool itself can be the cause, particularly if it is poorly installed.

Eliminating Chatter

Separate from these ideas already presented, chatter can also be eliminated by the way machining operations are selected prior to boring. For example, if an external tool removes the bulk of material by roughing, it may also leave a thin stock left for boring. Thin stock should always be of concern, as it is another cause of chatter. Rearranging the order of operations (bore first, then turn) may eliminate the chatter. Cutting depth has also been known to cause chatter, and reducing it may help to eliminate chatter. Keep in mind that cutting depth works together with spindle speeds and feedrates, so they both have to be considered

A Reader for Programmers

as well. Although the initial tendency to eliminate chatter is to reduce speeds and feeds, often the exact opposite is more likely to be successful.

There are at least two other factors that should be considered when encountering chatter, particularly on CNC lathes. One is to change the cutting insert. Inserts come with different types of built-in chip breakers. A positive type chip breaker will have a positive influence on eliminating chatter in many situations. The other factor that may help in eliminating chatter is the tool nose radius. For most lathe work, nose radius of 0.8 mm (0.0321 inches) is common. A smaller radius will result in less pressure during cutting, but presents challenges of its own, such as possible changes of cutting feedrates (in the program) and various offset settings (at the machine).

Although most suggestions presented here apply specifically to a CNC lathe, for milling operations, the basic principles apply equally. In both cases, try to find the actual cause before attempting the cure.

Lathe Cycles G70–G72 — Part 1

February 2009, updated February 2013

In this two-part essay, we will look at two very common roughing cycles and one finishing cycle for CNC turning and boring operations. In the first part, a general description and the cycle format will be presented. In the second part, general practices will be described, along with many do's and don'ts and other practical suggestions.

Material removal on CNC lathes is normally completed by using *multiple repetitive cycles*. There are three major repetitive cycles used for turning and boring:

- G70 — Finishing cycle, used to finish part shape generated by either G71 or G72

CNC Tips and Techniques

- G71 — Roughing cycle, used for turning or boring where the direction of cut is primarily horizontal
- G72 — Roughing cycle, used for turning or boring where the direction of cut is primarily vertical

One more cycle (not described here) is G73, which is also available but seldom used. This is a *pattern repeating cycle*, suitable for rough machining of various castings, forgings, and other suitable forms. Subsequent finishing also uses the G70 cycle.

The G71 and G72 Cycles

The G71 cycle is most commonly used for external or internal roughing (turning and boring). The roughing is mainly in the horizontal direction, whereas roughing with G72 is mainly in the vertical direction. In both cases, chamfers, tapers, and radiuses are typical parts of the contour. If you understand the format and usage of either cycle, you will learn the others more easily. There are two formats for these cycles: one format occupies only one program block, the other occupies two program blocks. The one-block method is older and applies to Fanuc controls 6-10-11-15. The two-block method is used for more recent Fanuc controls, such as 0-16-18-21, and later. Two-block formats are similar for both G71 and G72:

G71 U.. R..
G71 P.. Q.. U.. W.. F.. S..

G72 W.. R..
G72 P.. Q.. U.. W.. F.. S..

The first block uses address U or W for the roughing cut depth. *U-address* for G71 applies to the depth *per side* in X-direction; *W-address* for G72 applies to the depth in Z-direction in current dimensional units.

R-address is the amount of retract from each roughing pass.

A Reader for Programmers

This is nothing more than a retract clearance controlled by the programmer and is usually very small: 0.5–1.0 mm or 0.02–0.04".

The second block is identical to both cycles. The addresses P and Q always go together. The *address P* identifies the block number representing the tool position at the contour start, and the *address Q* identifies the block number representing tool position at the *contour end*. Both block numbers must be unique (not duplicated in the program).

Addresses U and W also work together; they represent the stock allowance left for subsequent finishing. The *U-address* is the amount of stock left on *diameter,* and the *W-address* is the amount of stock left of *faces*. For simultaneous XZ motions, the U and W stock allowances will be compounded by the control system. Note that U is a dimension on diameter — a 0.03" physical stock will be programmed as U0.06. A positive amount of U or W leaves stock on the positive side of the contour; a negative amount leaves the stock on the negative side of the contour, determined by the cutting direction. For the majority of work, the U-address will be positive for turning, and negative for boring. A W-address is typically positive for both types of operations, but it could be negative, for example, in some back-boring operations.

The *F-address* is the cutting feedrate and the *S-address* is the cutting speed, normally in G96 mode (constant cutting speed), programmed in *feet per minute* or *meters per minute*. If the speed has been programmed at the program beginning, it is not necessary to include it in the cycle call.

The G70 Cycle

Finishing cycle G70 is always programmed as a one-block command, with the mandatory P and Q, and optional F and S data:

G70 P.. Q.. F.. S..

CNC Tips and Techniques

Addresses P and Q are more often than not the same as those used for the roughing cycle.

Roughing and Finishing a Simple Part

Here is an example of one tool that will both rough and finish a simple part, using G71 and G70 cycles in imperial units:

```
N1 G20 T0100
N2 G96 S400 M03
N3 G00 G42 X6.2 Z0.1 T0101 M08 (START POSITION FOR THE CYCLE)
N4 G71 U0.15 R0.025 (ROUGH CYCLE - FIRST BLOCK)
N5 G71 P6 Q13 U0.06 W0.005 F0.02 (ROUGH CYCLE - SECOND BLOCK)
N6 G00 X3.2 (CONTOUR START POSITION)
N7 G01 X3.5 Z-0.05 F0.007
N8 Z-1.5 F0.012
N9 X4.0 Z-2.5
N10 Z-3.5 R0.375 (AUTO CORNER ROUNDING)
N11 X5.9
N12 U0.2 W-0.1 F0.007
N13 G00 U0.2 (CONTOUR END POSITION)
N14 G70 P6 Q13 S500 (FINISHING CYCLE)
N15 G40 G00 X5.0 Z3.0 T0100
N16 M30
%
```

In the second half of this essay, we look into more details of these three cycles and identify possible problem causing issues and their solutions.

A Reader for Programmers

 ## Lathe Cycles G70-G72 — Part 2
March 2009, updated February 2013

In this second essay of two, we look at the roughing and finishing cycles for turning and boring in more detail. When programming the G71/G72 lathe cycles, some general comments can be useful. Following the suggestions presented here may prevent some common programming difficulties:

Preventing Programming Difficulties

- In the two-block format, both blocks have to contain the G71 or G72 command — neither is a modal command.
- The P-address is always the block number where the contour starts (first point).
- The Q-address is always the block number where the contour ends (last point).
- Watch the P and Q numbers as they relate to existing block numbers. Change of block numbers, such as re-sequencing, could influence the P and Q numbers. I know of only one CNC-oriented editing software (NCPlot™ from www.ncplot.com) that will automatically renumber P and Q references, if the block numbers have been changed. Also, make sure that no block number is duplicated in the program
- Feedrates programmed between the P and Q blocks are for finishing cycle G70 only.
- The start position for the cycle is defined as the last XZ coordinate before the G71/G72 is called.
- For safety reasons, use the same start point for roughing with G71/G72 and finishing with G70, unless a special application is necessary.
- The cycle starts and ends at the initial start point, and the return to this point is automatic (do not return to the start point within the P-Q block range).
- The start point in the X-axis determines the actual depth of the first roughing pass in G71. For external cutting, increasing the start point diameter will decrease the actual depth of the first

CNC Tips and Techniques

cut; decreasing the start point diameter will increase the actual depth of the first cut.
- A G41/G42 tool nose radius offset must be entered before G71/G72 is called, and must be cancelled after the G71/G72 cycle is completed. Do not use G41/G42 within the P-Q range. A tool nose radius offset is ineffective for G71/G72 roughing cycles.
- For the two-block entry, the depth of cut is measured per side with a decimal point. For example, 0.1" depth will be written as U0.1 for G71 cycle and W0.1 for G72 cycle
- For the one block entry, the depth of cut is measured per side in the units of the smallest increment, without a decimal point. For example, 0.1" depth will be written as D1000.
- Changing the cutting depth also influences the last diameter depth, for example, to eliminate an extremely small last cut.
- The U-amount is the amount of stock left on diameter for finishing. Positive or negative value indicates on which side of the diameter the stock is to be left. Typically, positive stock is left on all external diameters whereas negative stock is left on internal diameters.
- The W-amount is the stock left on faces and shoulders. Most tools do not have a very large lead angle, and leaving a large W-amount may cause problems. Leave only 0.075–0.125 mm (0.003"–0.005") on vertical faces or shoulders
- The main cutting direction for G71 cycle is horizontal (usually towards the chuck), for G72 cycle it is vertical (usually towards the center line). Both cycles share all basic principles listed

Problems in Rough Cutting

Two common problems in rough cutting relate to the first and last cut. As mentioned above, there is a solution for both problems. Sometimes the first rough cut in G71 cycle is too shallow or too deep. It is possible to control the actual size of the first cut by changing the X-position of the start point. These three motions take place from any position to the position where the first cut is made:

A Reader for Programmers

1. The tool moves to the start point (wear offset applied during the motion).
2. G71 cycle block is executed and U/W offset is applied during the actual motion.
3. The depth is applied during actual motion.

Example

Geometry offset X-7.95 Z-6.63 (this is for reference only and is irrelevant for the actual calculation). Start from machine zero:

N3 G00 X4.0 Z0.1
N4 G71 U0.1 R0.03 (0.1 DEPTH OF CUT)
N5 G71 P.. Q.. U0.06 W0.005 D1000 F.. (0.06 STOCK ON DIAMETER)

From X7.95 to X4.0... tool is at X4.0 absolute
U-stock added to X4.0 + 0.06... tool is at X4.06 absolute
From X4.06 subtract double depth of cut... tool is at 4.06 - 2 x 0.1 = **X3.86**

The diameter of the first cut will be 3.86 inches. If the wear offset is in effect, it will be included in the calculation as well.

There is a lot more to multiple repetitive cycles than meets the eye. Understand them well, including all the details that are not so obvious.

Figure A6-1 in Appendix 6 summarizes the drawing and the program for this essay.

CNC Tips and Techniques

Limitations in Threading
April 2009, updated February 2013

Programming a single point thread is very simple, considering that virtually all threading operations use a special multiple repetitive cycle G76. This is a parametric type of a cycle that occupies only two blocks of the CNC program (older controls use a one-block format). This cycle was described in the essay G76: *Two Formats, One Cycle*. This essay focuses on problems that may occur because of certain limitations in threading, regardless of what programming method is used (long hand or cycle).

Threading problems in this category belong to two limitations:

- Limitation of spindle speed (rpm)
- Limitation of cutting feedrate

When evaluating a threading problem, pull out the machine manual and check what is the actual maximum programmable range of spindle speed and cutting feedrate. For the purpose of this essay, let's use these ranges as illustrative examples:

- Minimum spindle speed *125 rpm*, maximum spindle speed *3000 rpm*
- Minimum cutting feedrate *250 ipm* in X-axis and *500 ipm* in Z-axis

Pitch vs. Lead

One very important distinction in threading is the difference between the pitch and the lead of a thread. Much too often, the word *pitch* is used where lead is the correct term. *Lead* is defined as the distance the tool will advance along an axis in one spindle revolution, whereas *Pitch* is defined as the distance between corresponding points of two adjacent threads. *Lead* equals *Pitch* only for single start threads. Cutting feedrate for threading is always equivalent to the thread lead. For any thread, the feedrate is equivalent to:

A Reader for Programmers

Feedrate = Lead = Pitch * Number of starts

The G97 Command

Also important in threading is using the G97 spindle speed command, not the constant cutting speed G96. Having established the background, what exactly are the possible limitations in threading?

Take a reasonable *1300 rpm* as programmed spindle speed for a *16 TPI* thread with four starts. This speed had been proven for a similar single start thread and presented no problems.

Comparing Feedrates and Spindle Speeds

In order to find out if the CNC lathe can handle cutting the thread at this spindle speed, we have to compare the respective feedrates per minute:

Maximum cutting feedrate of the machine is *250 ipm* (always use the lowest rating of all axes), so the effective cutting feedrate will be *0.25 * 1300 = 325 ipm*.

For this particular thread, the *1300 rpm* is excessive, even when the same spindle speed is used with no problems for the similar thread with a single start. The thread lead is a fixed value based on the drawing and cannot be changed. That leaves only the spindle speed, which has to be *reduced*. You may guess, you may use various trial-and-error methods, but the best way is to establish the maximum speed by calculation:

MAX rpm = MAX ipm / Programmed feedrate per revolution (lead of thread)

In this example, the maximum spindle speed S will be:

MAX rpm (S) = 250 ipm / 0.25 feedrate = 1000 rpm

Always program the actual cutting speed S a little lower than the calculated maximum spindle speed, in order to avoid

CNC Tips and Techniques

possible spindle fluctuation. In this case, *900* to *950 rpm* should be a good range. In the program for this example, the block may be written as *G97 S950 M03*.

This explanation has shown a solution to a problem where the spindle speed was *too high* for the given thread. What about the other extreme — where the spindle speed is *not slow enough*? Is such a situation possible? Although rare in most cases, such a situation is certainly possible and is also more likely in multi-start threading. Consider the same example of *16 TPI*, but with *48 starts*, not just four. In certain industries, super-multiple start threads are the norm. First, calculate the threading feedrate per one revolution, the thread *lead*:

Feedrate = 1 / 16 * 48 = 3.0 inches per revolution (!)

Next, calculate the maximum spindle speed S in revolutions per minute:

Required spindle speed (S) = 250 / 3 = 83.333 = 83 rpm

If the resulting spindle speed is higher than the minimum spindle speed of the CNC lathe, the thread can be machined without any difficulties. This is not the case for the example as illustrated — the specifications of the CNC lathe indicate *125 rpm* as the minimum spindle speed. That means the calculated *83 rpm* is not suitable for precision machining of this particular thread. Although an excessive spindle speed can be reduced, insufficient spindle speed cannot be increased.

The two examples presented in this essay are certainly not an everyday issue in most machine shops. However, they do show that in extreme threading cases some special decisions have to be made

A Reader for Programmers

 ## Programming a Long Thread
May 2009, updated February 2013

In the essay *Limitations in Threading*, the main focus was on limitations in single point threading caused by spindle speed that is either too high or too low. In this essay, we look at the effect of threading feedrate, particularly how it applies to long threads.

Long Threads

What exactly is a long thread may be a subject for some discussion. For the purposes of this article, a long thread is a thread with a ratio of rather small diameter to long threading length. In most everyday threading operations, you will work with threads that have a length relatively short respective to their diameter. These threads, which do not require any physical support at the machine, generally do not present any problems in programming or machining. Machining long threads does have a few challenges, however. Long threads often require suitable support devices, at least a tailstock for basic support, but often a follower rest or a steady rest as well, in order to prevent deflection of the part.

Many CNC machine tool builders provide lathes specially designed for machining long parts, such as shafts and tubular stock. These lathes (often of the flat bed type) are also capable of threading over their entire Z-axis travel. From the CNC programmer's perspective, a possible problem lies with the feedrate for certain *imperial* threads. Feedrate for threading is always the thread lead, and no rounding is necessary for metric threads. Some imperial threads have to use rounding for feedrate calculation, when the threads per inch (TPI) are converted to the thread lead (cutting feedrate).

Sample Program

As an example of a possible problem of feedrate rounding,

CNC Tips and Techniques

consider a single start, standard form 60 degree thread 6.0-12, over the threading length of 70 inches (actual use of the thread is not important for the illustration):

- Thread diameter = 6.0 inches
- Thread per inch (TPI) = 12
- Length of thread = 70 inches

Assuming a suitable tooling and part setup, the thread cutting program is easy:

N51 T0500
N52 G97 S550 M03 (SPINDLE SPEED 550 RPM)
N53 G00 X6.25 Z0.4 T0505 M08 (START POINT CLEARANCE)
N54 G76 P011060 Q005 R0.002 (G76 - BLOCK 1)
N55 G76 X5.8978 Z-70.0 R0 P0511 Q0120 F0.0833 (G76 - BLOCK 2 - 0.0833 IPR)
N56 G00 X12.0 Z5.0 T0500   (TOOL CHANGE CLEARANCE)
N57 M01

All program data look reasonable, but let's focus on the programmed feedrate of F0.0833. For twelve threads per inch, the pitch is 1/12 = 0.08333333, which is also the thread lead and programmed feedrate. Now, consider the feedrate actually used in the program, F0.0833. It has been *rounded to four decimal places,* as is customary for imperial dimensions.

Inevitably, the rounding has brought in a certain amount of inaccuracy. Over one inch length, the error is:

1 - 12 * 0.0833 = 0.0004 inches

In most cases, this error can be absorbed by the thread tolerances. Now, consider the same error over a significant thread length, such as our example of 70 inches:

70 * 0.0004 = 0.0280 inches — this is a significant error

A Reader for Programmers

Because the 0.0833 amount was the correct rounding to four decimal places, any other rounding would make the accumulative error even worse. The solution to this problem is quite simple — just change the programmed feedrate of F0.0833 to a new feedrate of *F0.083333*, from four to six decimal places. Fanuc-type controls allow this method, for the exact purpose of minimizing the accumulative error for long threads. With **F0.083333**, the error over 70 inches will be:

1 - 12 * 0.083333 = 0.000004 inches * 70 = 0.00028, which is less than three tenths.

Older Fanuc and similar controls could not use F-addresses with six decimal places, but the control provided E-addresses, specifically for threading, which did allow six decimal places. The six decimal place accuracy is not required for threads that have lead divisible into four or fewer decimal places, for example 16 TPI = 1 / 16 = 0.062500.

The G76 threading cycle itself will be discussed in two other essays looking at basics and details.

Threading with G76 Cycle — Basics
June 2009, updated February 2013

Threading on CNC lathes generally falls into two categories: it either works well, or it brings many problems. Of course these are extremes, but there is a general reasoning that if threading problems do happen, they are not always easy to resolve. This causes a production delay to be avoided altogether, or at least minimized. The actual cause of the problem has to be found first, before a possible treatment can be applied. Ideally, the threading program itself should anticipate such difficulties and should be perfect. Although the intent is good, the reality is

CNC Tips and Techniques

that threading is an operation that uses the weakest tool with the heaviest feedrates applied. Having problems in threading is almost to be expected, because they are rather hard to predict.

The G32 Command

It is always desirable that any problem encountered in production is resolved as soon as possible. On the machine side, if threading does not work as it should, the CNC operator's options are somewhat limited. The method of programming makes a big difference. This view is certainly true with the so-called *long-hand* threading method, where each threading motion is written as a separate program block. A typical example of this method is the threading method that uses the G32 command (or G33 on some machines). In this case, the operator cannot do anything significant during machining to improve the threading passes because all motions are fully embedded in the part program. In order to improve the thread quality, the design of the program itself has to be changed, virtually from the ground up. This process is time consuming and does not guarantee a better thread.

Although the G32 method of thread programming can be used in special situations where the programmer needs absolute control over the threading process, the vast majority of threading on CNC lathes today is done by using a very powerful *multiple repetitive cycle*, using a threading specific cycle G76. Comparing this cycle with another popular cycle, G71/G72 for turning and boring, G76 has many more parametric settings that can be adjusted at the machine, during production, if necessary. This flexibility is what makes the cycle not only popular, but also very effective for lead time.

The G76 Cycle

The first thing you should understand is that there are *two* formats of the G76 cycle. The original one is a single block program entry; it applies to older controls such as Fanuc 10 and 11,

A Reader for Programmers

and a few others. Modern controls use a two-block format that offers much more flexibility in individual settings. This cycle greatly benefits CNC programmerd during program development, yet it benefits CNC operators even more because it offers many features that can be changed at the CNC machine in a very short time.

In the current two-block G76 format, there are parameters of the cycle that require careful scrutiny. Let's start with the structure of the G76 cycle itself:

G76 P.. Q.. R..
G76 X.. Z.. R.. P.. Q.. F..

The main reason Fanuc has implemented the two-block format was to include more settings options at the machine. Due to the assignment of various letters (addresses) available to the program development, some letters have to be used twice, for totally different purpose. So the P-Q-R has a different meaning in the first block than the same letters P-Q-R have in the second block. Note that the G76 is *not* a modal command and must be repeated in both blocks.

Now, to the cycle format itself. Rather than explaining what the individual cycle entries represent (that comes later), let's look at what CNC operators might require during actual thread cutting at the machine. If the threading process works well, no changes are necessary. If problems do occur, they appear mainly in two categories: usually a *premature tool wear* and a *poor thread finish*. There is no single common cure for these problems because their causes depend on factors too numerous to mention. The main benefit of the two-block G76 cycle is the number of changes the CNC operator can make on-the-fly at the machine, causing a minimal disruption of the production process. These are not just changes to a single setting, but often changes affecting several settings simultaneously to achieve the main objective of a quality thread. Experimenting with the G76 settings is often the best way — sometimes the only way — to solve any threading problem.

Although the general description of individual cycle entries is available from numerous sources, the details are often obscure and contain no useful details for practical applications.

Understanding the basic background of possible threading problems is essential to their solution. In a separate essay, we will look at the G76 cycle structure from the perspective of the CNC operator and evaluate individual cycle parameters in more detail. Look for continuation of this discussion in the essay *Threading with G76 Cycle — Details;* where the many format details of the G76 threading cycle will be discussed.

Threading with G76 Cycle — Details
July 2009

The essay *Theading with G76 Cycle —* Basics covered the basics of threading on CNC lathes, using the G76 multiple repetitive cycle. When a program containing G76 cycle reaches the CNC operator, all threading data are contained in two blocks. Older controls use only one block, but the two-block format is far more popular and will be described here in more detail.

Two-Block Format

As an example, a standard external thread of 12 threads per inch (12 TPI) will be made on a three-inch diameter. Its length will be 2.5 inches. The programmer will provide the following data:

G76 P010060 Q0040 R0.002
G76 X2.8978 Z-2.5 R0 P0511 Q0120 F0.083333

In the first block, the *P010060* is a six-digit entry in three

A Reader for Programmers

pairs, in the format of *Pxxyyzz,* where **P01***yyzz* is the *number of finishing* cuts in the range of 01 to 99. In this case, only one finishing cut is required. In reality, this is the number of spring passes. Feel free to use two or three, particularly for harder materials. The second pair **Pxx00***zz* is the *number of leads* for gradual pull-off, sometimes incorrectly called chamfering. Typically, this pair will be 00, if the thread ends in a groove, recess, or similar open space. For threads that end in solid material, the setting should be 10 (=1.0) or more, which will provide a smoother angular exit from the thread. Note that the range is from 0.1 to 9.9 — and without the decimal point, 1.0 lead is programmed as 10. Also note that the amount of lead will shorten the full-length thread, and extending the Z-amount in the second block may be necessary. Finally, the third pair **P***xxyy***60** is commonly used for 60-degree V-threads — imperial or metric. *This angle of tool tip* can use only six values: 00, 29, 30, 55, 60, and 80. This setting controls the infeed angle of the tool. Therefore, if you change 60 to 55, for example, the infeed angle changes, while the thread retains its 60-degree shape, because that is the physical angle of the insert.

In both blocks, there are addresses P and Q. They do not take decimal points and must be entered in the proper format, for example, 0.02" will be 0200, 0.15 mm will be 150. The Q-address in the first block is the *minimum threading depth.* In the example, Q0040 indicates 0.0040 inches. As the threading tool cuts deeper, the cycle automatically decreases the depth to maintain the same chip amount removal. If the minimum depth is calculated below the Q-setting, the Q-amount will take over and the depth does not decrease any further. Note that both P and Q have different meaning in the first and second block!

The last address of the first block is R, which is the fixed amount for *finishing allowance.* In the example, the finishing allowance is 0.002 inches. This is the depth of the last threading pass, and is usually smaller than the P-amount in the same block.

In the second block, the X-address is the *final diameter* of

CNC Tips and Techniques

thread, calculated from a common formula *0.613/TPI or 0.613*Pitch*. For 12 TPI, the depth is 0.613/12 = 0.0511, so the final thread diameter is 3.0–2*0.0511=2.8978. For internal threads, the X-address is the nominal thread size, but the bore size has to be calculated from a similar formula using *0.541/TPI or 0.541*Pitch*. The Z-address indicates the *Z-position* of the tool when it retracts from the thread. The R-address is the *difference between the end of thread and the start of thread*, measured per side. For straight threads, the difference is zero, programmed as R0 and can be omitted. For typical tapered threads, the R-amount will be *negative for external threads and positive for internal threads*. The P-address in the second block is the *height of the thread*, as calculated above — 0.0511, programmed as P0511. The Q-address is the amount of the *first thread depth* measured per side. Q0120 sets the depth to 0.0120 inches. The threading feedrate is always the thread lead, never the pitch. For single start threads, the lead is the same as pitch. In imperial units, the threading feedrate can be up to six decimal places. For metric threads, there is no practical use for extended number of decimal places.

Typically, the CNC operator can control the thread quality by considering changes to all data in the first block, and the Q-amount in the second block. Changes to other data are much less frequent.

Additional Points

In closing, keep in mind a few additional points. For staggered threading (changing sides of the thread during cutting), you have to use Fanuc 15 threading mode that incorporates a single block entry with four options specified by the address P. Generally, the feedhold button is ineffective during threading, but some controls allow it by forcing the thread to withdraw gradually. The XZ start point of the thread is also important, as it determines the position of the tool after each threading pass.

Threading with the G76 cycle is very efficient and can be used for majority of threads. On the other hand, always keep in mind that G32 long hand threading is always there for those very special cases.

 ## Feedrate Adjustment on Arcs
August 2009, updated February 2013

In general, for CNC work, the cutting feedrate is determined by the CNC programmers and is based on established machining practices. This decision typically includes the part material and its machinability, tool type and diameter, and depth and width of cut, as well as overall setup rigidity and other factors. More often than not, the other factors do not include special attention to feedrates along arcs. For the majority of jobs, the cutting feedrate for circular interpolation is identical to the feedrate for linear interpolation and causes no problems.

When a cutter makes a linear toolpath, the cutter center is offset by the cutter radius. The linear path generated by the cutter center is either longer or shorter than the actual part edge and is always parallel to it. In this case, there is no need to consider feedrate adjustment at all. When the same cutter makes a motion along an arc, the circular path of the cutter center is also parallel to the part edge, in the form of a concentric arc. In most cases, the surface finish quality of the linear path and the circular path are the same and nothing needs to be done.

A decrease in surface finish quality may occur in some cases, where the cutter radius is large in relation to a small arc of the part. A solution to this problem may be as simple as adjusting the cutting feedrate either up or down from the calculated normal feedrate. An adjusted arc feedrate is not required in every program. If the cutter center toolpath is close to the part drawing contour, no adjustment is needed. On the other hand, when a large diameter cutter is used to contour a small outside radius, a problem that affects the surface finish may occur. In this case, the tool center path generates a much longer arc than the one in the drawing. In a similar situation, if a large cutter diameter is used for an inside arc, the equidistant path will be much shorter than the original arc.

- An outside arc of a toolpath is longer than the arc in the drawing.
- An inside arc of a toolpath is shorter than the arc in the drawing.

The Fundamental Rule of Feedrate Adjustments for Arcs

Based on these two definitions, the fundamental rule of feedrate adjustment for arcs is that the programmed linear feedrate is increased for outside arcs and decreased for inside arcs.

Suppose a cutting feedrate of 18.0 inches per minute (F18.0) gives excellent results for linear motions, but noticeably poorer results on small arcs. The cutter diameter used is 1.5 inch, the small outside arc in question has a radius of 0.4 inches. In this case, the cutting feedrate for the outside arc has to be increased, using the following formula:

$$FO = F * (R + r) / R$$

where FO is the adjusted feedrate for outside arcs, F is the linear feedrate, R is the drawing arc radius, and r is the cutter radius. Using the formula above, the change in feedrate from F18.0 will result in a heavier feedrate of F27.6 (ipm) along the outside arc.

A similar example for an inside arc will require reduction of the linear feedrate. Suppose linear feedrate is 15 inches per minute (F15.0), the inside arc has a radius of 0.75 inches, and the cutter diameter is 0.5 inches. In this case, the cutting feedrate for the inside arc has to be decreased, using the following formula:

$$FI = F * (R - r) / R$$

where FI is the adjusted feedrate for inside arcs, F is the linear feedrate, R is the drawing arc radius, and r is the cutter radius.

A Reader for Programmers

Using the above formula, the change in feedrate from F15.0 will result in a much lighter feedrate of F10.0 (ipm) along the inside arc.

The two formulas provide the means to find the adjusted feedrate along an arc, which is mathematically equivalent to the linear feedrate. Both formulas are recommended for external or internal contouring only, not for rough machining in solid material.

In lathe programming, there is no reason to distinguish between feedrates for linear and circular tool motions, regardless of the radius size. Tool nose radius is usually small, averaging 0.0313 inches (or 0.8 mm) and the equidistant toolpath is very close to the programmed toolpath.

Keep in mind that the formulas provided here (like many other formulas) should serve as guidelines only and should be used as such. Common sense combined with practical experience should always prevail.

Knurling on CNC Lathes
September 2009, updated February 2013

As an operation, knurling is as old as manual lathe machining. Its main purpose is to make formed indentations in the material that suit various purposes. When it comes to programming a knurling operation on CNC lathes there are four basic motions:

- Rapid to a start position
- Feed in for the length of knurl
- Retract from part diameter
- Return to the start point or a tool change position

This essay looks at several issues important for successful knurling on CNC lathes.

CNC Tips and Techniques

Knurling Patterns

The knurled indentations in the material surface are caused by a rolling action of the knurling tool while it is moving along the horizontal machine axis. Depending on the knurling tool, these indentations can have either a straight, angular, or diamond (cross) pattern. An angular pattern can be left hand or right hand with different angles. Knurling is generally not a metal cutting action, but a metal forming action caused by force — not all lathes are suited for knurling.

Knurling tools especially developed for CNC lathe machining are the tools recommended. Some designs are suitable for knurling to a square shoulder with minimum clearance. Pay special attention to the knurling wheel itself and to its diameter and pitch. A typical drawing will specify the type of knurl used on the machined surface as one of three patterns:

- Straight knurl   parallel to work
- Angular knurl   typically at 30° or 45°
- Diamond knurl   also 30° or 45° (known as diagonal, cross, or checkered pattern)

Each of these types has a specific purpose in practice. The straight knurl is used mainly for joining two cylindrical objects, for example, a pin and a hole. The diagonal type is to facilitate easier handling, such as better hand grip on rotary knobs or handles.

Knurling tools have one or two wheels. Using one left hand and one right hand wheel in the same holder and opposite to each other, a diamond pattern can be cut. Knurling tools designed for CNC lathes have floating (self-centering) positioning built-in, to provide an accurate alignment with the part. A common designation of the pitch size is *TPI* (teeth per inch). The pitch of a knurl is also important in feedrate selection. There are three groups of knurl pitch: coarse, medium, and fine.

Knurling is a final or finishing machining operation, usually on an external part diameter, although internal knurling is also possible. That means all other machining operations have to be completed beforehand. Because knurling is a lathe type of machining, the part has to be turned to the specified diameter before the knurling operation can be applied.

Knurling Tool Motions

Tools for knurling have to be selected carefully, as they can only be used within a certain range of work diameters. CNC programmers are responsible for programming the tool motions, selecting the spindle speed and the cutting feedrate. For small threads, an additional support may be required, such as a tailstock.

The first of the four motions is to the start point; this point should be above the knurl, and about 0.100" (2.5 mm) into the part (Z-0.1 or Z-2.5 if the front face is Z0). The second motion is a feedrate motion, and the knurl will move to the cutting depth. Once the tool has reached the cutting diameter, it continues in the feedrate mode towards the end of the knurl — this is the third motion. Once the operation is completed, the tool leaves the part at rapid rate along the X-axis —this is the fourth motion. Having a generous supply of coolant, and teeth free of any slivers or burrs of metal are both critical. If the knurling length is smaller than the knurling wheel width, the tool can just plunge in and leave the diameter immediately. CNC operators can use various offset settings to fine tune the knurled depth. The knurling depth can also be calculated based on the circular pitch in terms of teeth per inch (TPI) but this is a bit more involved subject for a short essay.

It is not unusual to do some experimentation at the machine. Spindle speed is normally calculated for the material being machined. The plunging feedrate should be fairly aggressive, and when knurling is completed, the tool should be retracted right away. A very small angular tilt of the knurling tool often helps to relieve back pressure.

CNC Tips and Techniques

Imperfections

If the knurl is not perfect, the chances are that either it is too deep or too shallow. A knurl too deep produces burrs and nicks on the part surface. A knurl too shallow produces noticeable flats on the part surface. A short summary may provide some clues as to the likely causes:

- Slivers, burrs and nicks   Probable excessive cutting depth
- Lower than expected tool life   Lubrication problems, hard material cutting
- Insufficient depth   Wrong knurl or program data

Knurling can be an easy operation or a tough one. If you have never knurled, run a few tests first to prevent scrapping the part that matters.

Programming a Full Circle
October 2009, updated February 2013

When you program arcs, it is not unusual to see part programs using the R-radius instead of the traditional I and J vectors. For example, **G02 X.. Y.. R..** has largely replaced **G02 X.. Y.. I.. J..**

There is no doubt that the radius designation simplifies programming because no calculations for the radius are necessary. I/J vectors, on the other hand, have to be calculated as individual distances from the arc start point to the arc center, measured along each axis. What many programmers may not realize is that the direct R-radius has some limitations. Up to and including 180° arc sweep angle, the radius is programmed as a positive value. A radius that has a sweep angle of more than 180° but less than 360° is programmed with the radius having a negative value. That leaves an arc with a sweep angle of 360° degrees: a full circle.

Programming a full circle means programming a toolpath

A Reader for Programmers

that will machine a 360° arc. There are various methods of achieving this objective:

> Using the *quadrant* method (*four* blocks required)
> Using the *semicircle* method (*two* blocks required)
> Using the *one block* method (*one* block required)

In the standard XY plane G17, both I/J vectors and direct radius R can be used for the quadrant and semi-circle methods, but only I/J vectors can be used for the one-block method. Many CNC programmers use I/J vectors exclusively for their compatibility with many control systems. The start point for the circular motion is usually a quadrant point, and 0° (three o'clock) is commonly used. Another common position is at 12 o'clock (90°). A CW or CCW cutting direction is selected by the type of milling. Climb milling is often the preferred method for CNC machining.

For the following examples, X0Y0 is at the circle center; the cutting motion will start at 0° as P1, in CW direction (G02), with the radius of 20 mm. Only basic concepts are shown.

Quadrant Method

The method of dividing a circle into four quadrants is the original way of programming. Practically, it offers no advantage in modern programming and is included for comparison with the other two methods. Using this method, a full circle is programmed in four blocks — one block for each quadrant (90° arc sweep). A block for each arc must never cross the axis at 0°, 90°, 180°, or 270°. Either the I/J vector or the R method can be used.

```
N10 G90 G00 X20.0 Y0 (P1) (== USING I AND J VECTORS ==)
...
G02 X0 Y-20.0 I-20.0 J0 (P2)
X-20.0 Y0 I0 J20.0 (P3)
X0 Y20.0 I20.0 J0 (P4)
X20.0 Y0 I0 J-20.0 (P1)
G00 ...
```

CNC Tips and Techniques

N20 G90 G00 X20.0 Y0 (P1) (== USING DIRECT RADIUS R ==)
...
G02 X0 Y-20.0 R20.0 (P2)
X-20.0 Y0 R20.0 (P3)
X0 Y20.0 R20.0 (P4)
X20.0 Y0 R20.0 (P1)
G00 ...

Semicircle Method

A semicircle is one half of a circle or 180° arc sweep. For a full circle programming using this particular method, two half arcs are necessary.
N30 G90 G00 X20.0 Y0 (P1) (== USING I AND J VECTORS ==)
...
G02 X-20.0 Y0 I-20.0 J0 (P2)
X20.0 Y0 I20.0 J0 (P1)
G00 ...

N40 G90 G00 X20.0 Y0 (P1) (== USING DIRECT RADIUS R ==)
...
G02 X-20.0 Y0 R20.0   (P2)
X20.0 Y0 R20.0 (P1)
G00 ...

What is interesting about the semicircle method is that it is often used as the default for CAM-generated part programs. The main reason is that in order to output R-radius, the four-block method is not necessary and the one-block method cannot use R-radius at all. Of course, the output can be changed by the user through post-processor customization.

A Reader for Programmers

One-Block Method

The one-block method is the shortest method of programming a full circle. In order to program a 360° arc sweep, the direct radius R method cannot be used.

N50 G90 G00 X20.0 Y0 (P1) (== USING I AND J VECTORS ==)
...
G02 I-20.0 J0 (P1)
G00 ...

N60 G90 G00 X20.0 Y0 (P1) (== USING DIRECT RADIUS R ==)
...
(=== RADIUS R CANNOT BE USED ===)
...
G00 ...

Ignoring the quadrant method, which of the remaining two should you use? If you insist on using the R-radius, you have to use the semi-circle method, but the surface finish may have small marks in two places. The one-block method should produce a better surface finish, as there is only one-place contact. My personal preference is to use I/J vectors for their compatibility with virtually all controls.

 **Peck Drilling — Watch the Q**
November 2009, updated February 2013

As the name of the title operation suggests, peck drilling is a method of drilling with interruptions. Rather than one single drilling motion, peck drilling describes a drilling operation where the drill does not reach the programmed Z-depth in one cut, but in a series of cuts. This method is very useful for deep holes, so the coolant can reach the drill point with ease, cooling

CNC Tips and Techniques

the cutting edges while also allowing the chips to be flushed out, at least as much as possible.

Virtually all CNC machining centers support two programming commands (fixed cycles) for peck drilling:

- G83 = Deep hole drilling
- G73 = High speed deep hole drilling

Take these descriptions as guidance only — a hole drilled in a tough material can be shallow and still benefit from this method of machining. High-speed peck drilling is a bit of an exaggeration, and some manuals actually call this G73 cycle *chip breaking cycle,* which is more descriptive. The *high speed* refers to the retract motion of G73 when compared to G83.

Programming Format

Both G83 and G73 are common fixed cycles. Except the cycle number, their format is identical:

- G98/G99 G83 X.. Y.. R.. Z.. Q.. F..
- G98/G99 G73 X.. Y.. R.. Z.. Q.. F..

In both cases, the amount of Q is the *depth of each peck.* The difference between the two cycles is in the actual cutting, particularly in the retract motion after each peck:

In the G83 cycle, the tool will retract to the R-level after each peck. In the G73 cycle, the tool will retract only by a certain minimum (~0.5 mm or 0.02") after each peck.

In both cases, the retract motion from the final depth will be either to the initial level (in G98 mode) or to the R-level (in G99 mode).

Application Notes

The G83 cycle can be used for peck drilling of most holes, whereas the G73 cycle is better suited for holes that need chip

A Reader for Programmers

breaking, regardless of their depth. CNC operators often find that the Q-depth programmed is either too small or too large. Although the actual change of Q-depth at the machine is a simple edit, some knowledge about how the cycle works is helpful. Peck drilling always start from the R-level (R-address = feed plane) and ends at the specified Z-depth (Z-address). This observation is an important one for those times when the desired number of pecks is to be programmed and the Q-depth has to be calculated.

Note that the Q-depth will be equal for each peck, with the possible exception of the last peck, because the tool (drill) will never exceed the programmed drilling depth (Z-depth). For example:

G99 G83 X300.0 Y250.0 R2.0 Z-26.5 Q8.0 F250.0

Number of pecks =
Total travel between R and Z / Q (always rounded up)
Total travel = (2 + 26.5) / 8 = 28.5 / 8 = 3.5625
Number of pecks = 4 pecks

Most controls do not allow programming the actual number of pecks desired. In such cases, when you know that so many pecks will do the job efficiently, you still have to calculate the Q-depth. Here is a very interesting example I will use to illustrate the point. A drawing is not necessary, the numbers speak for themselves. The objective is to calculate the Q-depth as the *minimum depth* that guarantees *three pecks* with a *4.75*-mm diameter drill. Here is the G83 cycle; calculate the Q-depth:

G99 G83 X175.0 Y200.0 R3.0 Z-24.925 Q.. F200.0

Normally, part of the calculation would also include the Z-depth, including the drill point length, but our concern is the Q-depth only.

CNC Tips and Techniques

Because the Z-axis start position is at Z3.0 (programmed as R3.0 in the fixed cycle), the total travel length of the tool will be 27.925. The number of required pecks is three, *always* calculated over the total distance to travel, in our case, 27.925 mm.

The result of 27.925/3 is 9.308333 as the minimum Q-depth amount. Because rounding to three decimal places is required for metric units, the Q-depth must be rounded up and must be programmed as **Q9.309** (or higher) in the fixed cycle. Because the requirement is for the *minimum* Q-depth, Q9.309 must be used in the cycle. Take care in rounding! Correct mathematical rounding of 9.308333 to three decimal places is 9.308, but that does not meet the requirement of always rounding *upwards* for peck drilling. The minimum correct Q-amount will be Q9.309 and used in the cycle.

If you did use the minimum Q-depth as Q9.308, you would get 9.308 x 3 pecks = 27.924 depth, one micron (!) short — that is, 0.00003937 inches. I have intentionally used a metric example, taking advantage of a micron as the smallest unit a CNC machining center can accept. The result will be three pecks of 9.308 and the fourth peck of exactly one micron! By changing the Q-depth to 9.309, there will be two pecks of 9.309 and one peck of 9.307. So, the final answer is that the minimum Q-depth is 9.309 mm. As you see, even one micron does make a difference in CNC work.

Appendices

A Reader for Programmers

Appendix 1
June 2008

Interpreting a CNC Program

The program example did not define the next tool (T10) as a drill, but there are two clues:

1. The spot drill is commonly used for positioning accuracy. It is also used to make a 45-degree chamfer on the subsequent hole, as is the case here.
2. Bloc N6 contains the chamfer description.

Here is the block N6, as programmed:

N6 G99 G82 R2.0 Z-2.6 P200 F200.0 (0.35x45 CHAMFER)

The challenge question was:
Can you tell the diameter of the drill that follows the spot drill?

Here is the explanation and the answer:
In order to machine a 45-degree chamfer, the spot drill used must have a 90-degree tool tip angle (see Figure A1-1). The next item of importance is the Z-depth of Z-2.6. Because a triangle with two 45-degree angles also contains two sides of equal length, it follows that the depth of 2.6 is one side. The other side, also 2.6 mm, is one half of the chamfer diameter (distance from the centerline to the chamfer end):

Chamfer diameter = 2 x 2.6 = 5.2 mm

Given that the chamfer is 0.35 x 45 on each side of the centerline, we subtract 2 x 0.35 = 0.7 from the chamfer diameter, to get the following drill diameter:

5.2 − 0.7 = 4.5 diameter

which is the correct answer.

*****

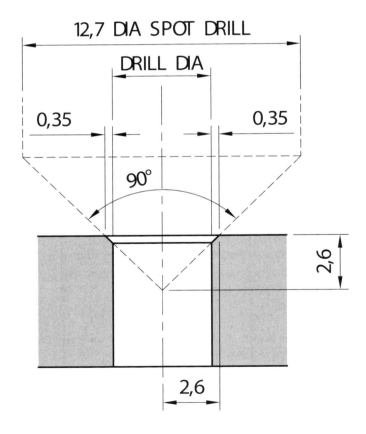

A Reader for Programmers

 Appendix 2
July 2008

Default Settings in Macros

The following program accompanies Figure A2-1.

```
(MAIN PROGRAM - EXAMPLE OF USAGE)
N1 G21
N2 G90 G00 G54 X0 Y0 S1200 M03
N3 G43 Z10.0 H01 M08
N4 G99 G81 R2.0 Z-15.9 F225.0 L0 (K0)
N5 G65 P8500 X50.0 Y37.5 D49.0 H6 A10.0 S1 (MACRO CALL)
N6 G80 Z10.0 M09
N7 G28 Z10.0 M05
N8 M30


O8500 (=== BOLT HOLE CIRCLE MACRO =======)
(== VARIABLE DEFINITIONS ================)
(X = #24 = X-CENTER LOCATION - DEFAULT=0 )
(Y = #25 = Y-CENTER LOCATION - DEFAULT=0 )
(D = #7  = BOLT CIRCLE DIAMETER         )
(H = #11 = NUMBER OF HOLES - RANGE 2-48 )
(A = #1  = FIRST HOLE ANGLE  - DEFAULT=0 )
(S = #19 = FIRST HOLE NUMBER - DEFAULT=1 )
(== ERROR TRAPPING SECTION ==============)
IF[#7 EQ #0] GOTO9101
IF[#7 LE 0] GOTO9102
IF[#11 EQ #0] GOTO9103
IF[#11 NE FUP[#11]] GOTO9104
IF[[ABS[#11] LT 2]] OR [[ABS[#11] GT 48]] GOTO9105
(== DEFAULTS ===========================)
IF[#24 NE #0] GOTO10
 #24 = 0
N10 IF[#25 NE #0] GOTO20
 #25 = 0
N20 IF[#1 NE #0] GOTO30
 #1 = 0
N30 IF[#19 NE #0] GOTO40
```

```
#19 = 1
N40 IF[#19 NE FUP[#19]] GOTO9104
IF[#19 LT 1] GOTO9106
IF[#19 GT #11] GOTO9107
#10 = #4003
(== MAIN MACRO BODY START ===============)
 #7 = #7/2
WHILE[#19 LE #11] DO1
 #30 = [#19-1]*360/#11+#1
 G90 X[COS[#30]*#7+#24] Y[SIN[#30]*#7+#25]
 #19 = #19+1
END1
(== MAIN MACRO BODY END =================)
GOTO9999
(== ERROR MESSAGES =====================)
N9101 #3000=101 (BOLT DIAMETER MISSING)
N9102 #3000=102 (CANNOT BE ZERO OR NEGATIVE)
N9103 #3000=103 (NUMBER OF HOLES MISSING)
N9104 #3000=104 (HOLES DATA MUST BE INTEGER)
N9105 #3000=105 (RANGE OF HOLES IS 2 TO 48)
N9106 #3000=106 (START HOLE IS 1 OR HIGHER)
N9107 #3000=107 (START HOLE NUMBER TOO HIGH)
(== MACRO END ==========================)
N9999 G#10
M99
%
```

A Reader for Programmers

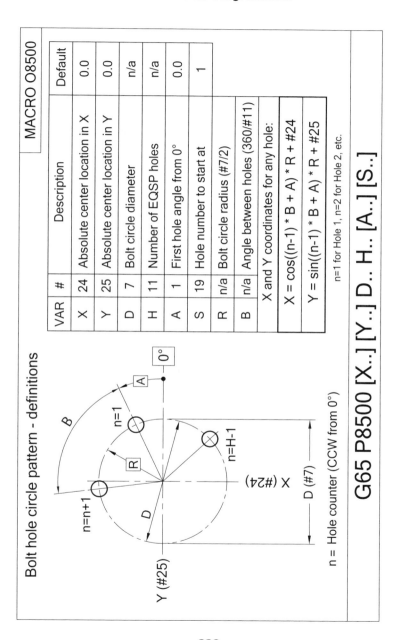

Bolt hole circle pattern - definitions

n = Hole counter (CCW from 0°)

MACRO O8500

VAR	#	Description	Default
X	24	Absolute center location in X	0.0
Y	25	Absolute center location in Y	0.0
D	7	Bolt circle diameter	n/a
H	11	Number of EQSP holes	n/a
A	1	First hole angle from 0°	0.0
S	19	Hole number to start at	1
R	n/a	Bolt circle radius (#7/2)	
B	n/a	Angle between holes (360/#11)	

X and Y coordinates for any hole:

$X = \cos((n-1) * B + A) * R + \#24$

$Y = \sin((n-1) * B + A) * R + \#25$

n=1 for Hole 1, n=2 for Hole 2, etc.

G65 P8500 [X..] [Y..] D.. H.. [A..] [S..]

CNC Tips and Techniques

Appendix 3
August 2008

Create Your Own G-Code

The following program accompanies Figure A3-1.

```
O9013 (G13 CIRCULAR MILLING CYCLE — CLIMB MILLING)
#31 = ABS[#3]/2 (POSITIVE RADIUS OF THE CIRCULAR POCK-
ET)
#11 = #4001 (STORE CURRENT G-CODE GROUP 01 — MOTION
COMMANDS)
#13 = #4003 (STORE CURRENT G-CODE GROUP 03 —
ABSOLUTE/INCREMENTAL)
#32 = #31-#[2000+#7] (TRUE POCKET RADIUS MACHINED)
IF [#32 LE 0] GOTO998 (ERROR IF RADIUS OFFSET VALUE IS
TOO LARGE)
#33 = #32/2 (CALCULATED LEAD-IN/LEAD-OUT RADIUS)
G91 G03 X#32 I#33 J0 F[#9/2] (MOTION 1 — LEAD-IN ARC TOOL-
PATH)
I-#32 F#9 (MOTION 2 - FULL CIRCULAR POCKET TOOLPATH)
X-#32 I-#33 J0 F[#9/2] (MOTION 3 — LEAD-OUT ARC TOOLPATH)
G#11 G#13 (RESTORE ORIGINAL G-CODES OF GROUPS 1 AND
3)
GOTO999 (BYPASS ERROR MESSAGE)
N998 #3000 = 13 (OFFSET TOO LARGE) (REFERS TO ACTUAL
OFFSET SETTING)
N999 M99 (MACRO END)
%
```

A Reader for Programmers

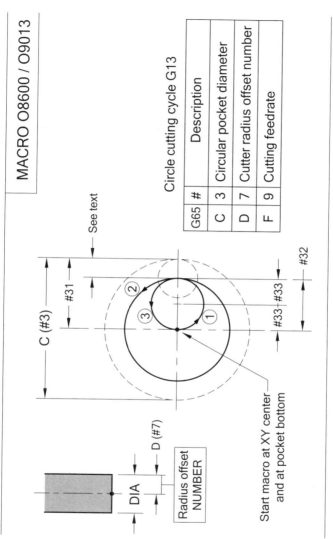

MACRO O8600 / O9013

Circle cutting cycle G13

G65 #	Description
C 3	Circular pocket diameter
D 7	Cutter radius offset number
F 9	Cutting feedrate

G13 C.. D.. F..

CNC Tips and Techniques

Appendix 4
September 2008

Scaling Option

The following program accompanies Figure A4-1.

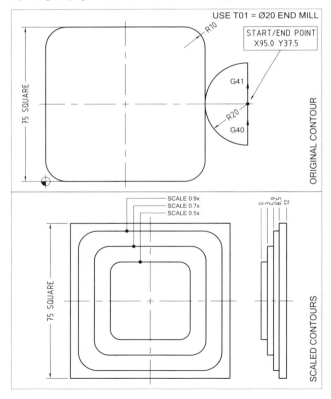

A Reader for Programmers

Program Comments and Listing

The center of scaling is the part middle point at X37.5 Y37.5. The key to a successful solution is to place the common toolpath into a subprogram, with one important feature - the tool **must return to the original start point**, after the scaling has been turned off - see block N115. It appears that the tool is at that position already - see X95.0 in block N112 and Y37.5 in block N113. That is *not* the case when scaling is in effect.

Look at both the program and the subprogram carefully - their structure is very important.

```
(SCALING.NC)
(MAIN PROGRAM)

(T01 = 20 MM DIA END MILL)
N1  G21
N2  G50                                         (SCALING OFF)
N3  G17 G40 G80 T01
N4  M06
N5  G90 G54 G00 X95.0 Y37.5 S2500 M03           (CONTOUR START POINT)
N6  G43 Z3.0 H01 M08
N7  G01 Z-3.0 F300.0                            (DEPTH OF THE SMALLEST CONTOUR)
N8  G51 I37.5 J37.5 P0.5                        (SCALING DATA - 0.5 SCALE)
N9  M98 P7003                                   (SCALED CONTOUR CUTTING)
N10 G01 Z-5.9 F300.0                            (DEPTH OF THE MIDDLE CONTOUR)
N11 G51 I37.5 J37.5 P0.7                        (SCALING DATA - 0.7 SCALE)
N12 M98 P7003                                   (SCALED CONTOUR CUTTING)
N13 G01 Z-8.5 F300.0                            (DEPTH OF THE LARGEST CONTOUR)
N14 G51 I37.5 J37.5 P0.9                        (SCALING DATA - 0.9 SCALE)
N15 M98 P7003                                   (SCALED CONTOUR CUTTING)
N16 M09
N17 G28 Z3.0 M05
N18 G50                                         (SCALING OFF - INCLUDED FOR SAFETY)
N19 G00 X-50.0 Y250.0                           (PART CHANGE POSITION)
N20 M30
%

O7003 (SUBPROGRAM FOR SCALING EXERCISE)
(D51 = CUTTER RADIUS = 10 MM)
N101 G01 G41 Y57.5 D51                          (LEAD-IN LINE WITH G41)
N102 G03 X75.0 Y37.5 I0 J-20.0 F250.0           (LEAD-IN ARC)
N103 G01 Y10.0
N104 G02 X65.0 Y0 I-10.0 J0
N105 G01 X10.0
N106 G02 X0 Y10.0 I0 J10.0
N107 G01 Y65.0
N108 G02 X10.0 Y75.0 I10.0 J0
N109 G01 X65.0
N110 G02 X75.0 Y65.0 I0 J-10.0
N111 G01 Y37.5
N112 G03 X95.0 Y17.5 I20.0 J0                   (LEAD-OUT ARC)
N113 G00 G40 Y37.5 F500.0                       (LEAD-OUT LINE WITH G40)
N114 G50                                        (SCALING OFF)
N115 X95.0 Y37.5                                (RETURN TO ORIGINAL START - NOT SCALED)
N116 M99
%
```

CNC Tips and Techniques

Appendix 5
October 2008

Safety and CNC Programming

CNC AND SAFETY

OVERVIEW

Machining safety is paramount to the successful operation of any machinery, CNC equipment included. Most companies adhere to - and in many cases exceed - the occupational safety laws and regulations of the jurisdiction in which they are located. A recent poster campaign has identified the basic rule of safety:

> Always observe all safety rules

This short advertising statement is catchy, clever - and - very true. Safety is often associated with the actual machining of a part, in machine shop environment. Although that is true, a CNC programmer has a great responsibility to built safety features into every CNC program.

SAFETY IN PROGRAMMING

Here is a list of some most common safety rules the CNC programmer should consider for CNC machining centers (also adaptable to the majority of CNC lathes):

- Approach program development with safety in mind for all work
- Program approach towards a part in XY axes first, then move the Z-axis
- Program retract from a part in Z axis first, then move the XY axes
- Start every program with cancellations (so called status block or a safety block)
- Provide sufficient clearances (above part, below part, lead-in and lead-outs, etc.)
- Program reasonable speeds, feeds, depth of cuts, width of cuts, etc.
- Select safe tool change or part change position
- Select suitable tools for the job
- Program tool motions towards a solid support, such as the fixed jaw of a vise
- Be consistent in programming approach from one job to another
- If program fails, correct the REAL cause of a problem, not a perceived one

SAFETY IN MACHINING - GENERAL

Because of the nature of work, the CNC operator has a lot more safety related issues to consider. Here is a list of some most common safety rules the person operating the CNC machine must follow:

- Wear approved safety boots
- Wear approved safety glasses with protective side shields and approved safety helmet, if required
- Wear suitable clothing (tucked-in shirt, buttoned-up sleeves, no ties)
- Keep long hair under a net or tied up
- Remove watches, rings, bracelets, and similar jewelry before machine operation
- Keep floors clean and free of oil or other hazards
- Do not remove guards and protective devices
- Do not use rags or gloves around moving or rotating objects
- Do not leave objects on top of machines
- Mount every part in a secure way

A Reader for Programmers

CNC AND SAFETY

- Check fixtures and tools before they are used
- Watch for sharp edges and burrs - deburr all sharp edges from parts
- Use only sharp tools - dull tools may be dangerous
- Do not use tools that are outside of specifications (weight, length, diameter, ...)
- Leave two adjacent tool pockets empty, if this feature is supported for slightly heavier tools
- When running the first part of a batch, watch for program errors, drawing errors, setup/tooling errors, ...
- Watch for small parts that can 'fly' off during part or tool rotation (inserts, shims, screws, pins, ...)
- Do not let the chip conveyor to accumulate too many chips
- Stop all machine motions when measuring or inspecting parts
- Stop all machine motions when leaving the machine temporarily (break time, washroom, ...)
- Stop all machine power for maintenance
- Electrical or control maintenance should be done by authorized personnel
- Do not use a grinding machine near the CNC machine slides
- Do not use a welding equipment on CNC machine under power
- Watch for built-up pressure in air hoses
- Watch various fluids (coolant, lube, tramp oil, dielectric fluid, ...) - they may cause problems
- Read and follow instructional and operating manuals supplied with the machine
- Do not alter design or functionality of the machines or controls
- Do not operate a faulty machine
- Pranks and horseplay are dangerous around machinery - behave responsibly
- Know when to use the Emergency Switch and when not
- ... Use common sense

SAFETY IN MACHINING - CNC

In addition, some specific CNC oriented safety suggestions, before and during machining:

- Interpret the part program correctly
- Understand the purpose of the program
- Make sure you are working with the latest drawing version
- Use various machine and control features during the first part run (single block, dry run, overrides, etc.)
- Make sure the tools are set properly in the magazine and registered accordingly
- Check with the CNC programmer or an engineer, if in doubt

> OTHER RULES AND STANDARDS MAY BE SPECIFIC TO A PARTICULAR PLACE OF WORK, AND EVERY EMPLOYEE SHOULD BE FAMILIAR WITH SUCH RULES AND REGULATIONS

CNC Tips and Techniques

Appendix 6
March 2009

Lathe Cycles G70-G72 — Part 2

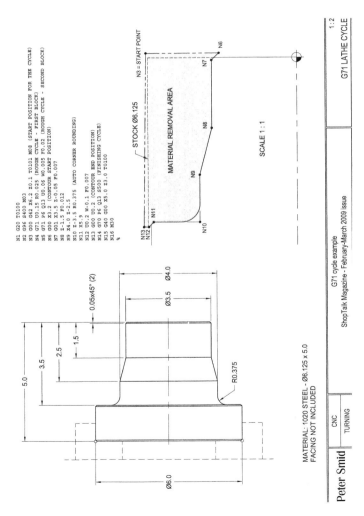

```
N1 G20 T0100
N2 G96 S400 M03
N3 G00 G42 X6.2 Z0.1 T0101 M08 (START POSITION FOR THE CYCLE)
N4 G71 U0.15 R0.025 (ROUGH CYCLE - FIRST BLOCK)
N5 G71 P6 Q13 U0.06 W0.005 F0.02 (ROUGH CYCLE - SECOND BLOCK)
N6 G00 X3.2 (CONTOUR START POSITION)
N7 G01 X3.5 Z-0.05 F0.007
N8 Z-1.5 F0.012
N9 X4.0 Z-2.5
N10 Z-3.5 R0.375 (AUTO CORNER ROUNDING)
N11 X5.9
N12 U0.2 W-0.1 F0.007
N13 G00 U0.1 (CONTOUR END POSITION)
N14 G70 P6 Q13 S500 (FINISHING CYCLE)
N15 G40 G00 X5.0 Z3.0 T0100
N16 M30
%
```

G71 cycle example
ShopTalk Magazine - February-March 2009 issue

G71 LATHE CYCLE

Peter Smid   CNC TURNING

INDEX

2D toolpaths 35
3D 19, 22, 35, 79, 129

absolute coordinates 107
absolute mode 52, 78, 90, 109
acceleration 62, 121, 163
accuracy 114
alarms 141
algebra 155
angle 108, 129
ANSI 29
arc motion 102
arcs 5, 107, 205–207, 210
argument 164
arithmetic 155
associated geometry 35-36
ATC 122–124, 126–127, 162
Autocad 3, 15, 38
automatic corners 43, 97–100
automatic tool change, see ATC
axial travel 119
axis inversion 84
axis motion 50, 72, 85
axis symmetry 84

batch 10, 180
blend radius 97
block delete 91
block numbers 44
block skip function 10–11, 23, 61, 91–93, 153
blocks 23–24, 46, 52, 54, 71, 76
bolt hole circles 49, 107, 109, 128, 164-166
boring, 3, 19, 20, 46, 100, 145, 149, 185, 187, 191
boundaries 19
breaks 97–100
brevity 20
burrs 97

CAD 29, 38
CAD/CAM software 12, 14–16, 22, 27
calculations 3–4, 28–30, 154–156
calculators 6, 156
CAM 19, 20, 33–39, 43, 150, 164
CAM machinist 12-16
CAM system 38–39
canned cycles 19
capacity 20, 22, 59
carbide 114, 122, 186
casting 170
Catia 38
C-axis 149
center 49
centerline 3, 145
certification 25
chamfer 97–100, 129, 188
changing programs 7
changing tools 8
chatter 185–187
chip breakers 187
chucks 5, 148, 186
circle 210–213
circular interpolation 102, 113
circular motion 101
clamping 70, 163
clearances 97, 174
climb milling 85

CNC Tips and Techniques

CNC lathes 43, 54, 72, 100, 145–151, 185, 207–210
CNC mills 96, 122
COM 112
comments 24, 25, 44–45
common offset 112
Compact II 19
comparing CNC 59–60
compression 63
cones 151–154
consistency 9, 137
constant surface speed 145
contours 5, 13, 35, 53, 158, 161
control 7, 59
control system 73, 127–130
conventional milling 85
conversion 46–48
coolant activity 70
coolants 176
coordinate rotation 95, 96, 129
coordinate system 49–51
corners 43, 49, 80, 96–100
costs 10–11
counter bore 167
counters 43
custom machine shops 14
custom macros 164, 167
customers 15–16
customization 36
cutter radius 87–88101, 102, 158, 167, 205,
cutting 24, 80, 92, 133
cutting depth 111, 174, 192
cutting speed 145–148, 185
cutting tools 178
cycles 19–21, 37
cylindrical interpolation 151

D 139–140

damping 186
data setting command 51–53
date 26
deceleration 63
decimal point 4
deep holes 57
default settings 164–166
deflection 151
depth 13, 67, 174, 185, 214
designers 18
diameter 54, 97, 125–126, 145, 163, 164
dimensional accuracy 9
dimensions 28–30, 49, 89–90, 129, 140, 158–160
direct drawing dimension 129
direct numerical control See DNC
direct parametric values 129
direction 85
DNC 21, 22, 35, 43, 45, 95
documentation 25–27
double block format 46–47
drawings 18, 27, 97, 129, 133, 157
drilling 3, 19, 35, 37, 57, 131, 145, 149, 213–216
dwell 161, 163
dwell time 17–18
DWG files 15, 38
DXF (drawing exchange files) 15, 38

EdgeCam 15, 94
editor 23
EDM machines 37
engineers 14, 18, 20–21, 49
errors 3–5, 9–10, 160, 172
Excel 156
EXT 22, 112

A Reader for Programmers

external offset 112

face cuts 93
Fadal 19, 25, 46, 57, 60, 88, 90, 96, 97, 107, 112, 116, 127, 145, 164, 167, 169, 171, 177, 188, 199–201
feeds 11, 24
feedrate 4, 17, 24, 47, 59, 62, 78, 81, 114, 119, 121, 162, 174, 185, 186, 191, 194–197, 205–207, 209
file transfer 35
fillet 97
finish contour 47
finishing 20, 48, 53, 152, 187, 189, 191, 203
first part 9, 11
fixed cycles 19, 43, 94, 96, 101, 102, 108, 130–133, 161
fixed-access ATC 122
fixture 179
flat machining 129
floating tap holders 62, 64, 67
flutes 122, 125
formulas 155, 206–207
Fortune 500 14
forward slash 93, 153
full circle 210–213
functions 70

G00 100
G01 78, 100, 161
G02 115, 161
G03 115, 161
G10 49, 51–53
G13 167-168
G15 108
G16 108
G17 100–102, 110
G18 102
G19 102
G20 140, 161
G21 78, 82, 140, 161
G22/G23 79
G25/G26 79
G27/G28/G29/G30 79–80, 90
G28 87–90
G31 80
G32 54–56, 72, 118, 120, 200
G33 54
G41/G42 85, 87–88, 110, 192
G43 73
G44 80
G50 146, 171–172
G51 171–172
G52 49–51, 108
G54 50, 52, 73, 110–112, 139
G55 51, 52
G60 81
G65 165, 167–169
G6x 80–81
G70 20, 116, 140, 152, 187, 189–190
G71 20, 21, 46–47, 116, 140, 152, 188–189, 191–193, 200
G72 116, 188–189, 191–193, 200
G73 57–58, 214
G74 63, 69
G76 20, 21, 46–48, 54–56, 116–118, 120, 194, 199-205
G81 19
G82 19, 108, 161
G83 19, 57-58, 214
G84 63, 69, 176
G90 52, 82,89
G91 52, 78, 82, 89
G92 54, 55, 120, 139
G96 145–146
G97 145-146, 195

233

CNC Tips and Techniques

gage line 126
G-codes 8, 78-81, 167-170
geometry 12–13, 35–37, 146, 155–156, 169
GibbsCam 15
graphic cards 35
grooving 37

H 139-140
Haas 60, 95
hardware 34–35
heat 151
helical interpolation 35, 113, 128
hobs 114
holes 56-59, 62, 76, 94, 100, 115, 128, 149, 164
homing 23
HSS 122

IGES (initial graphics exchange specification) 15
imperial units 140, 161
inches 28–29
incremental mode 52, 78, 82, 90, 109
inserts 122, 125, 187
inside arc 5
integrated software 37
intermediate point 89
interpolation 150–151
Inventor 38
ISO 9000+ certification 25, 29

job shops 14, 14

knurling 207–210

lathe control 55
lathe cycles 19–21, 46–48, 187–193
lead 119, 194–195, 197, 203
lead-in, lead-out 13
learning curve 12–13
left hand tapping 63
length 43–45, 126
linear interpolation 78, 113
live tooling 148–151
local coordinate system 49–51
location 23, 149
logic 4
long programs 22–24
long threads 197–199
longest tool method 74–75
lubrication 151

M00 23, 27, 70, 127, 140, 153
M01 23, 70, 91, 140
M03 70, 162
M04 70, 72, 162
M05 70, 72, 162
M06 70, 71, 127
M08 70, 93
M09 70
M19 141
M23 86–87
M30 70, 76, 140
M98 70, 76
M99 70, 76
machine functions 70
machine type 27
machine utilization 31
machine zero 79–80, 87–90, 171
machining 6, 9, 11
machining centers 96, 100, 113, 122, 125
machining cycles 19-21
machinist 12-13
Machinist Toolbox 156

A Reader for Programmers

macros 43, 57, 95, 129, 130, 164-167, 169-170, 176-178
manual data input see MDI
manual programming 6-8, 34, 37
mass substitution 44
Mastercam 6, 15, 33, 94
material removal 187
maximum tool diameter 125-126
MDI 8, 115, 123
measurements 151-154, 158-160
memory 22, 34
metric units 28-29, 140, 161
metric mode 78
M-functions 8, 61, 70-72, 86, 124, 140-141, 149, 162
milling 37, 82, 96, 113, 187
minimm threading depth 203
minimum increment 30
mirror images 84-87, 95
miscellaneous functions 70
Mitsubishi 95, 127
modular fixturing 11
mold work 22
multi-axis motions 44-45
multi-axis toolpaths 22
multi-machine support 37–38
multiple repetitive cycles 19-20, 43, 46, 187-188, 193
multi-process machining 148
multi-start threads 55, 119-121

NCPlot 95-96
negative sign 4
new parts 9-10
numbers 28-30

offset 4-5, 50, 51, 73, 80, 101, 102, 110-112, 139-140, 146, 157-160, 169, 170, 179, 192

Okuma 127, 164
one-block method 213
operators 3, 11, 16, 25, 30-33, 55, 94, 110, 125, 134-135, 151, 157, 160, 176, 178
optimizing 11, 17, 44
optional program stop 70, 91
origin 49
overhang 185

pallets 128
part catchers 128
part reversal 148
part zero 49-53, 73, 107, 110
peck drilling 213-216
peripheral speed 145
pipe thread 114
pitch 119, 194-195
planes 100-103
planning 10
plastic injection molds 170
pocketing 35, 100, 161
polar coordinates 96, 107-110, 128-129, 151
post-processors 7, 19, 36, 43
precision machining 151
preparatory command 54-55, 78-81, 167
preset method 73–74
process planners 10
production 9, 180
program 179
program end 70, 71
program errors 4
program generation 36
program length 131–133
program stop 23, 70, 71
program structure 81–84
program updating 17–19

235

CNC Tips and Techniques

program upgrading 16–18
program zero 49–53, 107, 110
programmer/operator 7, 30–33
programmers 6, 10–11, 13, 16, 19, 26, 30–33, 43, 46, 49, 54, 94, 110, 133–137, 151, 157, 170, 205
programming 6–13, 16–27,43–45, 69, 97–100, 115–116, 133–138, 150, 160–163, 173–175, 197–199, 210–213
Pythagorean Theorem 156

quadrant method 211–212
quality 114, 136–138, 205

radius 108, 188, 210
Random Access Memory (RAM) 34
random-access ATC 122–124
rapid motions 18, 59, 77, 78, 92
reaming 3, 145, 149,
records 25–27
rectangular coordinate system 107
repetition 130–133
resistance 62
restart 22
retract planes 24
retraction 54, 174
right hand tapping 63
right triangles 156
rigid tapping 62, 67–69, 128
rotary axes 128
rotation 70, 95, 129
rotational direction 145
rough cutting 192–193
roughing 20, 21, 47, 53, 187, 188. 191
round stock 152

rounding 4
rounding errors 172
rpm 145, 149, 194

safety 173–175, 191
safety block 82
scale 28
scaling 96, 129, 170–172
scrap 157–160
sections 23, 81–82
semicircle method 212
sequence 44
sequence numbers 23–24
sequence return 22
setups 5, 11, 27, 68, 73–75, 102–103, 125, 133, 148, 179
shaded models 94
shrinkage 129
Siemens 127
simulations 11, 94-96
single block format 46–47, 72
single point threading cycle 21, 54, 55, 145, 194
single-axis motions 44–45
skills 12, 18–19, 80, 91–93
slots 149
small shops 33–39
software 6, 20–21
software 20–21
SolidEdge 38
Solidworks 38
special instructions 27
speeds 11, 162, 174
spindle fluctuation 79
spindle rotation 70, 85, 114, 149
spindle speed 4, 8, 17, 47, 59. 62, 63, 69, 72, 145–148, 162, 176, 186, 194–196, 209
spot drilling 108, 130–131, 161

A Reader for Programmers

staggered threading 204
standard tapping 62–64, 67
start position 54
status block 174
steel 186
stock allowances 13, 47
stop 80, 91
stored stroke check 79
straight threads 55
stripping 176–177
subprograms 43, 70, 75–78, 96, 130, 132, 164, 167
sub-spindles 148
surface finish 205
synchronization 120
synchronized tapping 67
syntax 4

T03 152
tailstock 5, 163
tap drill sizes 176
TAPE 22
tapered threads 55
tapers 114, 117, 155, 188
tapping 3, 19, 62–64, 67–69, 80, 113, 131, 145, 149, 176–178
technical support 39
tension 63
tension-compression heads 62, 67
text editor 38
thread cutting 113, 128
thread length 117
thread milling 113–116, 128
thread stripping 176–177
threading 20, 21, 37, 46, 54–56, 119–121, 145, 194–196, 199–205
threading cycle 47–48, 55–56, 118
threads 62, 68, 113–116

threads per inch (TPI) 197–198, 202
three-axis center 100, 148
time 26
tolerances 18, 29
tool change 4, 59, 70–71, 84, 122–124, 162, 175
tool length 73–75, 80, 111, 126, 139
tool nose radius 4, 192
tool number 124, 139, 162
tool offsets 179
tooling 27, 122–127, 148–151
toolpath 35–36, 85, 94–96, 100, 128, 129, 133, 170–171, 179, 205
toolpath geometry 12, 38
touch-off method 74
training 6, 16, 39
transfer 148
translators 15–16
trial cuts 151–154, 180
trigonometry 3, 107, 156
turning 3–5, 20, 37, 46, 100, 145, 149, 187, 191
turrets 148
two-axis lathes 3, 148
two-block format 55, 202-204
type of operation 161

underfeeding 63–64
units 161
updating 17–19, 52
upgrading 17–18
user macros 164

vector 98
vendors 7, 20, 36, 39
verification 10
vertical lathes 95

237

CNC Tips and Techniques

vertical machining center 73, 149, 173
vibration 186

wear 151, 169
weight 126–127
wireframe simulation 94
work offset 50, 110–112

Yasnac 60, 127

zero 4, 44–45, 49–53